Passive Micromixers

Passive Micromixers

Special Issue Editors

Kwang-Yong Kim
Mubashshir A. Ansari
Arshad Afzal

MDPI • Basel • Beijing • Wuhan • Barcelona • Belgrade

MDPI

Special Issue Editors

Kwang-Yong Kim
Inha University
Korea

Mubashshir A. Ansari
Aligarh Muslim University
India

Arshad Afzal
Indian Institute of Technology Kanpur
India

Editorial Office
MDPI
St. Alban-Anlage 66
Basel, Switzerland

This is a reprint of articles from the Special Issue published online in the open access journal *Micromachines* (ISSN 2072-666X)) from 2017 to 2018 (available at: http://www.mdpi.com/journal/micromachines/special_issues/passive_micromixers)

For citation purposes, cite each article independently as indicated on the article page online and as indicated below:

LastName, A.A.; LastName, B.B.; LastName, C.C. Article Title. *Journal Name* **Year**, *Article Number*, Page Range.

ISBN 978-3-03897-007-1 (Pbk)
ISBN 978-3-03897-008-8 (PDF)

Contents

About the Special Issue Editors

Kwang-Yong Kim received his B.S. degree from Seoul National University in 1978, and his M.S. and Ph.D. degrees from the Korea Advanced Institute of Science and Technology (KAIST), Korea, in 1981 and 1987, respectively. He is currently a Dean of Engineering College and an Inha Fellow Professor in the Department of Mechanical Engineering of Inha University, Incheon, Korea. Professor Kim is also the current co-editor-in-chief of the International Journal of Fluid Machinery and Systems and the associate editor of ASME Journal of Fluids Engineering. He served as the editor-in-chief of the Transactions of Korean Society of Mechanical Engineers, the president of Korean Society for Fluid Machinery, and the chairman of the Asian Fluid Machinery Committee. He is also a fellow of the Korean Academy of Science and Technology, a member of National Academy of Engineering of Korea, a fellow of the American Society of Mechanical Engineers (ASME), an associate fellow of the American Institute of Aeronautics and Astronautics (AIAA), and a recipient of order of science and technology merit, "Doyak Medal" from Republic of Korea. He is interested in applications of the numerical optimization techniques using various surrogate models and computational fluid dynamics to the designs of fluid machinery, heat-transfer augmentation devices, micro heat sinks, micro mixers, coolant channels in nuclear reactors, etc. He has published about 370 peer-reviewed journal papers, and presented about 530 papers at international/domestic conferences.

Mubashshir A. Ansari is currently working as an Associate Professor in the Department of Mechanical Engineering, Aligarh Muslim University (AMU), a post he has held since May 2016. He received his B. Tech and M. Tech in Mechanical Engineering (Thermal and Fluid Flow) from AMU, Aligarh. After earning a PhD degree from Inha University, he worked as an Assistant Professor at Inha University for two years and as a postdoctoral researcher at Sogang University, Korea Advanced Institute of Science and technology (KAIST), Gwangju Institute of Science and Technology (GIST), South Korea for more than four years. Dr. Ansari has published a number of papers in reputed international journals and presented his research work at different conferences. His research area is in the field of MEMS/microfluidics, both numerical and experimental, mainly the design of micromixers, mixing of fluids at the microscale and the separation of particles using surface acoustics.

Arshad Afzal received his Ph.D. degree in the Thermal and Fluids Engineering Division, Department of Mechanical Engineering, Inha University, Republic of Korea in 2015. He received his M. Tech (Thermal Engineering) and B. Tech (Mechanical Engineering) degrees from Aligarh Muslim University in 2009 and 2006, respectively. Presently, he is working at the INSPIRE Faculty in the Department of Mechanical Engineering at the Indian Institute of Technology Kanpur, India. His research interests are mixing at microscales, turbulent flow and heat transfer in ribbed ducts, thermo-fluid design optimization, and surrogate modeling.

micromachines

MDPI

Review

Editorial for the Special Issue on Passive Micromixers

Arshad Afzal [1]**, Mubashshir A. Ansari** [2] **and Kwang-Yong Kim** [3,*]

1   Department of Mechanical Engineering, Indian Institute of Technology Kanpur, Kanpur,
    Uttar Pradesh 208016, India; aafzal@iitk.ac.in
2   Department of Mechanical Engineering, Aligarh Muslim University, Aligarh, Uttar Pradesh 202001, India;
    mub.ansari@yahoo.com
3   Department of Mechanical Engineering, Inha University, Incheon 22212, Korea
*   Correspondence: kykim@inha.ac.kr; Tel.: +82-032-860-7317

Received: 15 May 2018; Accepted: 17 May 2018; Published: 21 May 2018

Micromixers are important components of microfluidic systems, such as the lab-on-a-chip and micro-total analysis system (μ-TAS) designed for a large number of applications, including sample preparation and analysis, drug delivery, fast chemical reactions, and biological and chemical synthesis. Typically, micromixers operate from a very low to high Reynolds number range of 0.001–1000 in the laminar flow regime. Therefore, the random turbulent motions that are necessary to homogenize fluid samples are absent at the microscale. As a result, the mixing processes inside these microfluidic systems are diffusion-dominated and slow in addition to requiring a long channel length. Therefore, efficient mixing over a short channel length is a challenging task. Micromixers can be broadly classified as active (based on external energy/stimulus) and passive (based on geometrical modifications) micromixers. Among the two basic concepts, the passive micromixers do not rely on external mechanisms and offer the advantages of simple design and easy fabrication. This special issue focuses on presenting the development of efficient and robust designs of passive micromixers.

A total of 10 original research papers, which cover important aspects of micromixing technology, are published in the special issue. The important areas being focused upon are the numerical and experimental analyses of flow and mixing in different micromixers [1–5], optimization [6–8], and fabrication of micromixers [9]. In the review paper [10], recent developments in both active and passive micromixers are summarized and discussed.

Computational fluid dynamics (CFD) analysis of flow and mixing is seen as an important tool in designing passive micromixers in eight research articles [1–8]. Javaid et al. [1] proposed a serpentine-shaped micromixer with sinusoidal walls and conducted a CFD analysis using COMSOL Multiphysics software. The numerical results showed a good agreement with a previously published experiment, with improved mixing performance compared to a simple serpentine channel. A similar investigation of flow and mixing in serpentine channels with a non-rectangular cross-section was conducted by Clark et al. [2]. The results indicate enhancement in mixing performance using non-rectangular cross-section serpentine micromixers. Ansari et al. [3] performed both numerical and experimental analyses of a vortex micro T-mixer using ANSYS-CFX. The design is promising as it can efficiently increase the mixing for T-junction micromixers. Wang et al. [4] carried out CFD analysis of a herringbone passive micromixer for pressure-driven and electro-osmotic flow. A good mixing performance was observed at low flow rates for both flow situations. However, at a fixed flow rate, the mixing performance is superior in the case of electro-osmotic flow than pressure-driven flow. Okuducu and Aral [5] conducted a different study, which focused on the comparative evaluation of the finite volume method (FVM) and finite element method (FEM) to characterize numerical errors in mixing analysis. Using T-mixer geometry and two CFD tools, which were namely OpenFoam (FVM-based) and COMSOL Multiphysics (FEM-based), a detailed analysis based on mesh structure, orientation and flow parameters was presented. Using CFD and Taguchi statistical method, Solehati et al. [6] studied a T-junction micromixer with wavy walls to evaluate the effect of key design parameters on the mixing performance, pumping power and figure of merit (mixing index per unit pressure drop). The Taguchi method was found to be robust in determining the optimum combination of the design parameters. Raza et al. [7] carried out multi-objective optimization of a serpentine micromixer with crossing channels at low and high Reynolds numbers. Pareto-optimal fronts representing the trade-off between conflicting objectives, mixing index and pressure drop were obtained for both low and high Reynolds numbers. Guo et al. [8] demonstrated topology optimization based on Lagrangian mapping method for mixing problems that were dominated completely by the convection with negligible diffusion. The layout of the passive micromixers was determined by solving a topology optimization problem to minimize the mixing measurement.

Shan et al. [9] fabricated two types of micromixers with complex structures using the femtosecond laser wet etch (FMWE) technology inside fused silica. Using FMWE technology, the multi-microchannel mixers with high integration and uniformity for high-performance applications were realized, demonstrating the flexibility and universality of FLWE technology. Finally, the review paper by Cai et al. [10] summarized the recent advances in passive and active micromixers for microfluidic applications. In recently published articles, active micromixers have used pressure fields, electrical fields, acoustics, magnetic fields, and thermal fields as external energy sources to perturb the fluid. Furthermore, passive micromixers based on geometrical modifications using two-dimensional obstacles, unbalanced collisions, convergence–divergence structures or three-dimensional lamination and spiral structures were discussed.

We would like to take this opportunity to express our appreciation to all the authors for submitting their papers and contributing to the success of this special issue. We also want to thank all the reviewers for dedicating their time and helping to improve the quality of the submitted papers and professional staff at the Micromachines Editorial office for providing invaluable assistance.

References

1. Javaid, M.U.; Cheema, T.A.; Park, C.W. Analysis of Passive Mixing in a Serpentine Microchannel with Sinusoidal Side Walls. *Micromachines* **2018**, *9*, 8. [CrossRef]
2. Clark, J.; Kaufman, M.; Fodor, P.S. Mixing Enhancement in Serpentine Micromixers with a Non-Rectangular Cross-Section. *Micromachines* **2018**, *9*, 107. [CrossRef]
3. Ansari, M.A.; Kim, K.-Y.; Kim, S.M. Numerical and Experimental Study on Mixing Performances of Simple and Vortex Micro T-Mixers. *Micromachines* **2018**, *9*, 204. [CrossRef]
4. Wang, D.; Ba, D.; Liu, K.; Hao, M.; Gao, Y.; Wu, Z.; Mei, Q. A Numerical Research of Herringbone Passive Mixer at Low Reynold Number Regime. *Micromachines* **2017**, *8*, 325. [CrossRef]
5. Okuducu, M.B.; Aral, M.M. Performance Analysis and Numerical Evaluation of Mixing in 3-D T-Shape Passive Micromixers. *Micromachines* **2018**, *9*, 210. [CrossRef]
6. Solehati, N.; Bae, J.; Sasmito, A.P. Optimization of Wavy-Channel Micromixer Geometry Using Taguchi Method. *Micromachines* **2018**, *9*, 70. [CrossRef]
7. Raza, W.; Ma, S.-M.; Kim, K.-Y. Multi-Objective Optimizations of a Serpentine Micromixer with Crossing Channels at Low and High Reynolds Numbers. *Micromachines* **2018**, *9*, 110. [CrossRef]
8. Guo, Y.; Xu, Y.; Deng, Y.; Liu, Z. Topology Optimization of Passive Micromixers Based on Lagrangian Mapping Method. *Micromachines* **2018**, *9*, 137. [CrossRef]
9. Shan, C.; Chen, F.; Yang, Q.; Jiang, Z.; Hou, X. 3D Multi-Microchannel Helical Mixer Fabricated by Femtosecond Laser inside Fused Silica. *Micromachines* **2018**, *9*, 29. [CrossRef]
10. Cai, G.; Xue, L.; Zhang, H.; Lin, J. A Review on Micromixers. *Micromachines* **2017**, *8*, 274. [CrossRef]

micromachines

MDPI

Article

Analysis of Passive Mixing in a Serpentine Microchannel with Sinusoidal Side Walls

Muhammad Usman Javaid [1], Taqi Ahmad Cheema [2] and Cheol Woo Park [1,*]

[1] School of Mechanical Engineering, Kyungpook National University, 80 Daehak-ro, Bukgu, Daegu 41566, Korea; usmanjavaid90@gmail.com
[2] Department of Mechanical Engineering, Ghulam Ishaq Khan Institute of Engineering Sciences and Technology, Topi 23460, Khyber Pakhtoon Khwa, Pakistan; tacheema@giki.edu.pk
* Correspondence: chwoopark@knu.ac.kr; Tel.: +82-53-950-6550

Received: 22 November 2017; Accepted: 25 December 2017; Published: 28 December 2017

Abstract: Sample mixing is difficult in microfluidic devices because of laminar flow. Micromixers are designed to ensure the optimal use of miniaturized devices. The present study aims to design a chaotic-advection-based passive micromixer with enhanced mixing efficiency. A serpentine-shaped microchannel with sinusoidal side walls was designed, and three cases, with amplitude to wavelength (A/λ) ratios of 0.1, 0.15, and 0.2 were investigated. Numerical simulations were conducted using the Navier–Stokes equations, to determine the flow field. The flow was then coupled with the convection–diffusion equation to obtain the species concentration distribution. The mixing performance of sinusoidal walled channels was compared with that of a simple serpentine channel for Reynolds numbers ranging from 0.1 to 50. Secondary flows were observed at high Reynolds numbers that mixed the fluid streams. These flows were dominant in the proposed sinusoidal walled channels, thereby showing better mixing performance than the simple serpentine channel at similar or less mixing cost. Higher mixing efficiency was obtained by increasing the A/λ ratio.

Keywords: micromixer; chaotic advection; serpentine-shaped microchannel; mixing index

1. Introduction

Microfluidic devices use fluid flow at the submillimeter scale for applications in areas such as life sciences, analytical chemistry, and bioengineering. Small sample volume consumption, low cost, flexible and controlled operation, and high throughput make the use of microfluidic devices desirable [1]. Micro total analysis systems and microscale devices, which are employed for biochemical analyses and processes, such as protein folding, enzyme reactions, and drug delivery, require rapid mixing of reagents before a chemical reaction could occur [2]. Given the small characteristic dimensions at the microscale, the Reynolds number is low and the flow is laminar. Thus, without turbulent mixing, only molecular diffusion causes mixing, but is a slow process [3]. Therefore efficient mixing mechanisms are necessary to achieve the realized potential of lab-on-a-chip technologies.

Almost three decades ago, Ottino presented an overview of earlier work on mixing and chaotic advection, and comprehensive theory on kinematics and chaotic dynamics [4,5]. Following the interest of researchers in microfluidics, Ottino and Wiggins provided a review of mixing at microscale, and the mathematical foundations of chaotic mixing for design of efficient micromixers [6,7]. Past researchers have developed active and passive techniques to attain rapid mixing by influencing the flow to cause chaotic advection or to increase the contact area of fluid layers [8]. Active micromixers use external energy sources, such as pressure [9], acoustic [10–12], and electric field [13]. Active mixers such as the acoustic-based mixers with sharp edges [10–12] have been shown to achieve higher mixing efficiency than passive mixers, but they need an external energy source. By contrast, passive mixers use geometric characteristics to split, stretch, fold, and break the fluid streams, thereby enhancing mixing [14]. Although the fabrication of passive mixers is complex, the absence of an external driver and the ease of integration in microsystems provides them an edge over their active counterparts [8].

Previous studies on passive mixers have mainly relied on lamination- and chaotic advection-based designs to enhance mixing [14]. Buchegger et al. reported a multi-lamination mixer that used wedge-shaped vertical inlets, leading to a single horizontal channel, where efficient mixing of four streams occured [15]. Nimafar et al. proposed a mixer with H-shaped channels, for the splitting and recombination of two fluids. Experimental comparisons with T- and O-micromixers showed the superior performance of their device [16]. An experimental and numerical study on a crossing manifold micromixer has shown that the change in flow profile due to non-uniform momentum, along with convection and increasing interfacial area, improved mixing efficiency [17]. A three-layer split and recombination mixer achieved a high mixing efficiency because the design allowed for three times the surface area than a simple T-mixer could provide for diffusion to occur [18]. Kim et al. used chaotic advection and splitting and recombination in their serpentine laminating mixer. Three-dimensional serpentine mixing units produced chaotic advection, and F-shaped mixing units caused splitting and recombination [19].

Other strategies to enhance passive mixing utilize chaotic advection induced by geometric manipulations. Mengeaud et al. conducted an optical and numerical investigation on a zigzag microchannel. Molecular diffusion was the dominant factor for mixing at Reynolds numbers lower than 80, and secondary flows affected species mixing at high Reynolds numbers [20]. Hossain et al. used numerical investigation to compare the mixing performance of zigzag, curved, and square-wave-shaped channels and concluded that the square-wave-shaped channel exhibited better performance than the other two geometries [21]. Parsa and Hormozi investigated mixing in sinusoidal channels by varying the phase shift between side walls. High mixing indices were achieved for phase shifts of $\pi/2$ and $3\pi/2$ [22]. In another study, Parsa et al. investigated the effect of the amplitude to wavelength (A/λ) ratio of sinusoidal walls and observed the best performance at high A/λ ratios [23]. Afzal and Kim used coupling of pulsatile flow and sinusoidal walled convergent–divergent channel to achieve high mixing efficiency [24]. Asymmetric curvilinear microchannels can show better mixing performance than symmetric channels at Dean numbers (K) greater than 16.8, whereas symmetric channels achieve a higher mixing index than asymmetric curvilinear microchannel below the threshold value of K [25]. Fan et al. presented a study on the use of sharp corners in series to improve mixing efficiency [26]. Alam et al. presented a numerical study of straight and curved microchannels with circular obstacles that changed the flow pattern to achieve better performance than channels without obstacles [27].

Apart from achieving high mixing efficiency, micromixers should also process samples without damaging large biomolecules. Obstacle-based passive designs have high local strains and can damage biomolecules due to shear. Serpentine-shaped mixers can prevent this damage because, chaotic advection due to high local strains does not occur in these channels [28]. Therefore serpentine channels can be employed to achieve high mixing efficiency and damage-free processing of samples. Modifications, such as the use of non-aligned inputs [29] and three-dimensional serpentine geometries have been shown to further enhance the mixing performance of serpentine channels [28]. The study conducted by Alam and Kim showed that modifying the side walls of curved serpentine channels can also improve mixing performance [30]. They used rectangular grooves at specified locations and observed increased mixing as groove width expanded. Meanwhile groove depth only slightly affected the mixing index.

This paper reports the results of a numerical study on the mixing performance of a chaotic-advection-based serpentine microchannel with sinusoidal side walls. Sinusoidal walled channels have been shown to achieve better mixing performance than straight channels [22,23], however, the geometries used in the previous studies on sinusoidal walled channels were made by the modification of a straight channel. The present study presents a modified geometry that was designed using sinusoidal side walls in a serpentine channel to increase the secondary flow because of change in Dean number, and attain better mixing than a simple serpentine channel. The mixing performances of a simple serpentine channel and a serpentine channel with sinusoidal walls, with Reynolds numbers ranging from 0.1 to 50, were compared. A simple serpentine geometry similar to the square wave channel [21] and serpentine microchannel [29] reported by Hossain et al. was chosen to compare the mixing performance of proposed micromixer. Two additional cases of sinusoidal walled serpentine channel were considered by increasing the (A/λ) ratio of sinusoids. Micromixers with sinusoidal walls showed better mixing efficiency than a simple serpentine channel, and the increase in A/λ ratio further enhanced the performance.

2. Micromixer Design

The simple serpentine channel and serpentine channel with sinusoidal walls used in the present study are shown in Figure 1a,b, respectively. Both designs used two inlets, connected by a T-joint, that lead the fluids into the mixing channel. Before entering the serpentine-shaped part, fluids passed through a straight channel of 0.2 mm in length for both geometries. The inlets have a square cross section, and both planar devices have a width and depth of 0.1 mm. The sinusoidal walls shown in Figure 1b were generated using the following function:

$$y = A \sin\left(\frac{2\pi x}{\lambda}\right) \tag{1}$$

where A is the amplitude and λ is the wavelength of the sinusoids. Three cases of serpentine channels with sinusoidal walls with amplitudes of 0.02, 0.03, and 0.04 mm corresponding to A/λ ratios of 0.1, 0.15, and 0.2, respectively, were considered in this study. For the serpentine channel with sinusoidal walls, the sinusoids at the outer turns were joined using quadratic curves. The dimensions for both geometries used in the present study are shown in Figure 1.

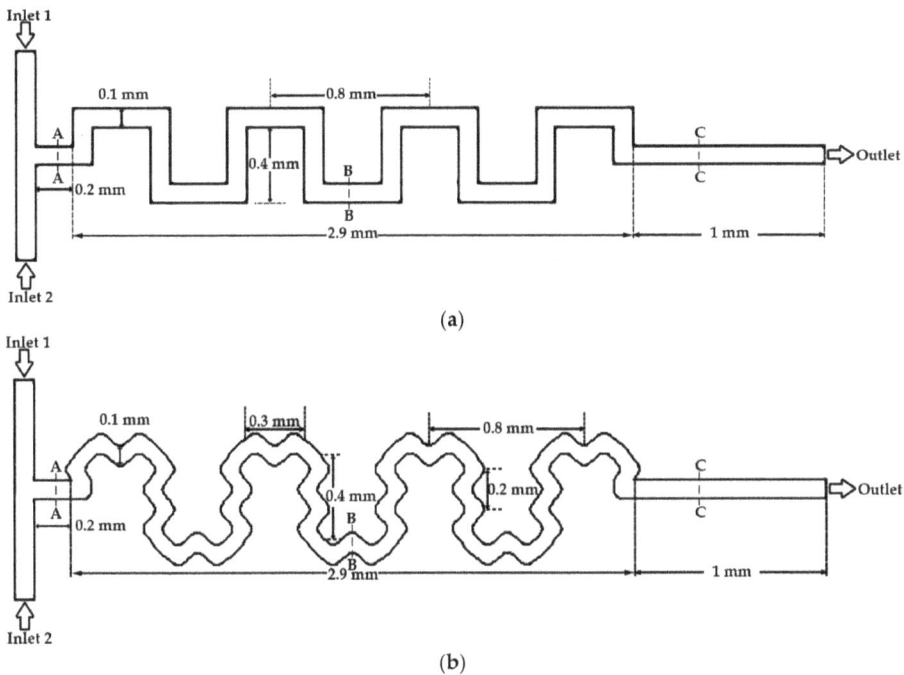

Figure 1. Schematics of microchannels. (**a**) Simple serpentine channel and (**b**) serpentine channel with sinusoidal walls.

3. Numerical Model

The single-phase, incompressible, and steady-state laminar flow in micromixers was solved for momentum and mass conservation using the Navier–Stokes equations and the continuity equation, respectively. The equations are expressed as follows:

$$\rho(u \cdot \nabla)u = \nabla \cdot [-pI + \mu(\nabla u + (\nabla u)^T)] \tag{2}$$

$$\rho \nabla \cdot (u) = 0 \tag{3}$$

Equation (2) represents momentum conservation in which ρ is the density of fluid (kg·m^{-3}), u is the velocity vector (m·s^{-1}) p is the pressure (Pa), I is the unit diagonal matrix, and μ is the dynamic viscosity of the fluid (kg·m^{-1}·s^{-1}). Equation (3) is the continuity equation. The solution of these equations yielded the velocity and pressure fields. The obtained velocity field was used to compute species concentration field using the convection–diffusion equation expressed as follows:

$$\nabla \cdot (-D\nabla c) + u \cdot \nabla c = R \tag{4}$$

where D is the diffusion coefficient (m^2·s^{-1}), c is the species concentration (mol·m^{-3}), and R is the reaction rate, which was assumed to be zero in this case.

4. Mixing Analysis

Methods based on striation thickness of fluid layers [31] and standard deviation of concentration [32] have been used in the past to characterize the mixing performance of micromixers. The formula employed in the present study to calculate mixing index (MI), based on the standard deviation of concentration is expressed as follows:

$$\text{MI} = 1 - \frac{\sigma}{\sigma_{\text{Max}}} \tag{5}$$

where σ is the standard deviation of species concentration in any given cross section and σ_{Max} is the standard deviation of the completely unmixed state. The value of the mixing index is 0 and 1 for the unmixed and fully mixed states, respectively. The standard deviation is expressed as follows:

$$\sigma = \sqrt{\frac{1}{N} \sum_{i=1}^{N} (c_i - c_m)^2} \tag{6}$$

where N is the number of sampling points, c_i is the mixing fraction at point i, and c_m is the optimal mixing fraction.

The power required to drive fluids through the channel should also be taken into account while characterizing the mixing performance of any design. Usually high mixing efficiency at increased flow rates results in higher power consumption [33]. So the cost of mixing should also be determined. This mixing cost is calculated in terms of pressure drop using the mixing index to pressure drop (MI/ΔP) ratio [34], or in terms of input power [35]. The formula used to calculate mixing cost (MC) in terms of input power is expressed as follows:

$$\text{MC} = \frac{Input\ Power}{\text{MI}} = \frac{\Delta P \cdot Q}{\text{MI}} \tag{7}$$

where ΔP is the pressure drop (Pa) across the channel and Q is the corresponding flow rate (m^3·s^{-1}).

5. Model Implementation

Water and a dilute dye solution were considered as working fluids, each with a density of 1000 kg·m^{-3} and a dynamic viscosity of 0.001 kg·m^{-1}·s^{-1}. The diffusion coefficient of 10^{-10} m^2·s^{-1}

was used for the dye solution in water. Any change in the physical properties of the fluid due to the presence of solute was ignored. To solve fluid flow, the no-slip boundary condition was set at the walls along with zero pressure at the outlet, and symmetry in the vertical direction. Velocity was used at the inlets, and simulations were conducted at various Reynolds numbers ranging from 0.1 to 50. Reynolds number is defined as follows:

$$Re = \frac{\rho U D_h}{\mu} \tag{8}$$

where Re is the Reynolds number, ρ is the fluid density (kg·m^{-3}), U is the fluid velocity (m·s^{-1}), D_h is the hydraulic diameter (m), and μ is the dynamic viscosity (kg·m^{-1}·s^{-1}) of the fluid. Inlet concentrations of 0 and 1 were used at the two inlets to compute the concentration field.

COMSOL Multiphysics (Version 5.3, COMSOL Inc., Burlington, MA, USA) was used for the simulations using the laminar flow and transport of diluted species interfaces. The domain was discretized using tetrahedral elements. Grid independence tests were carried out with different numbers of mesh elements (Figure 2). Finally to save the computational cost, 990,338 elements were used for the simple serpentine geometry. For three different cases of serpentine channels with sinusoidal walls, 710,957, 715,895, and 744,686 elements were used for the geometries with A/λ ratios of 0.1, 0.15, and 0.2, respectively. In numerical simulations, the discretization of convective terms for determining the concentration distribution causes numerical errors that result in the addition of numerical diffusion. The extent of numerical diffusion can be minimized using higher order discretization [36]. To reduce the extent of artificial diffusion, a higher order discretization was used in the present study. All simulations were conducted on a Windows 7 operated workstation with an Intel Xeon E5-2620 v3 2.4 GHz processor (Intel Corporation, Santa Clara, CA, USA) and 32 GB random access memory (RAM).

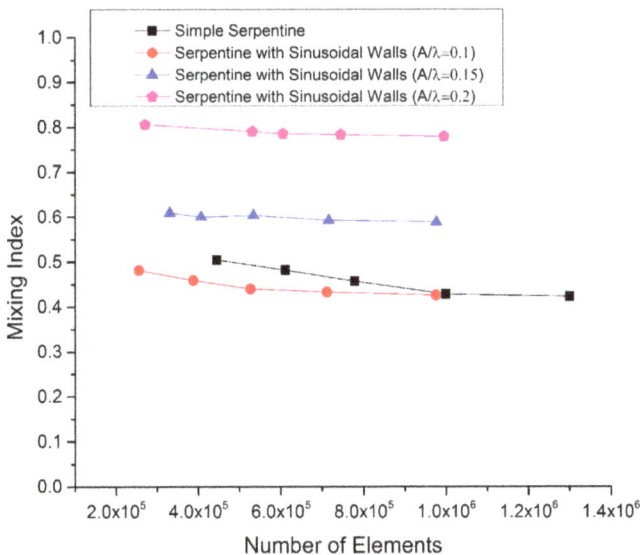

Figure 2. Mesh independence test at Reynolds number (Re) = 40.

6. Results and Discussion

The numerical model was validated by comparing its results with experimental results of Fan et al. [26]. Simulations were conducted using the reported geometric parameters and fluid properties. The results shown in Figure 3 validate the numerical model by showing close agreement of the simulation results with the experimental results.

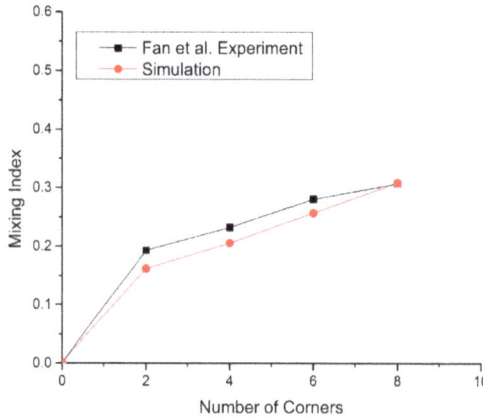

Figure 3. Comparison of the simulation and experimental results.

Four cases, namely, simple serpentine channel (Case 1), serpentine channel with sinusoidal walls and A/λ ratio of 0.1 (Case 2), serpentine channel with sinusoidal walls and A/λ ratio of 0.15 (Case 3), and serpentine channel with sinusoidal walls and A/λ ratio of 0.2 (Case 4), were considered in the present study. In curvilinear channels, the fluid motion towards the outer wall due to centrifugal forces creates velocity and pressure gradients that cause vortical flows. The magnitude of these secondary flows increases with the increase in Dean number (K) which is defined as follows:

$$K = \sqrt{\frac{D_h}{2r}} Re \qquad (9)$$

where D_h is the hydraulic diameter (m), Re is the Reynolds number, and r is the radius of curvature (m). The serpentine shape considered in the present study has sinusoidal walls as well as turns of the serpentine. Determination of radius of curvature and calculation of Dean number is difficult because of this geometric complexity. Equation (9) shows that for a geometry with fixed hydraulic diameter, the Dean number can be increased by increasing the Reynolds number or decreasing the radius. From this we can say that the Dean number in sinusoidal walled channels with a higher A/λ ratio would be greater because of increased curvature.

Streamline plots at Reynolds numbers of 0.1, 20, and 50 are shown in Figure 4 to understand the effect of secondary flows and the mixing phenomenon at different flow rates. An increase in Reynolds number will cause an increase in Dean number and as a result a change in streamlines trajectories for all cases can be seen because of increased secondary flow. At Reynolds number of 0.1, streamlines in all four cases move downstream of the channel with negligible path crossing (Figure 4a). Any mixing at low Reynolds numbers is dominated by diffusion because crossing of streams does not occur. Some crossing is observed at a Reynolds number of 20 (Figure 4b) and becomes dominant when the Reynolds number is increased to 50 (Figure 4c). Secondary flows start developing with the increase in the flow rate, thereby promoting fluid mixing. The mixing of streamlines also depends on channel geometry, as evident from the crossing of streamlines in different geometries at the same Reynolds number. This is because an increase in curvature also causes a rise in the Dean number. High mixing can be observed in the serpentine channel with sinusoidal walls compared to the simple serpentine channel because of more secondary flow, which becomes prominent as the amplitude of side walls increased. Flow separation also starts with the increase in the fluid velocity and amplitude of side walls. This phenomenon can be observed by the development of separation vortices, which are most effective in the sinusoidal walled channel with A/λ ratio of 0.2 (Figure 4c(iv)). This increased effect of secondary flows with the increase in curvature is consistent with previously reported results on curvilinear channels [22,23,25].

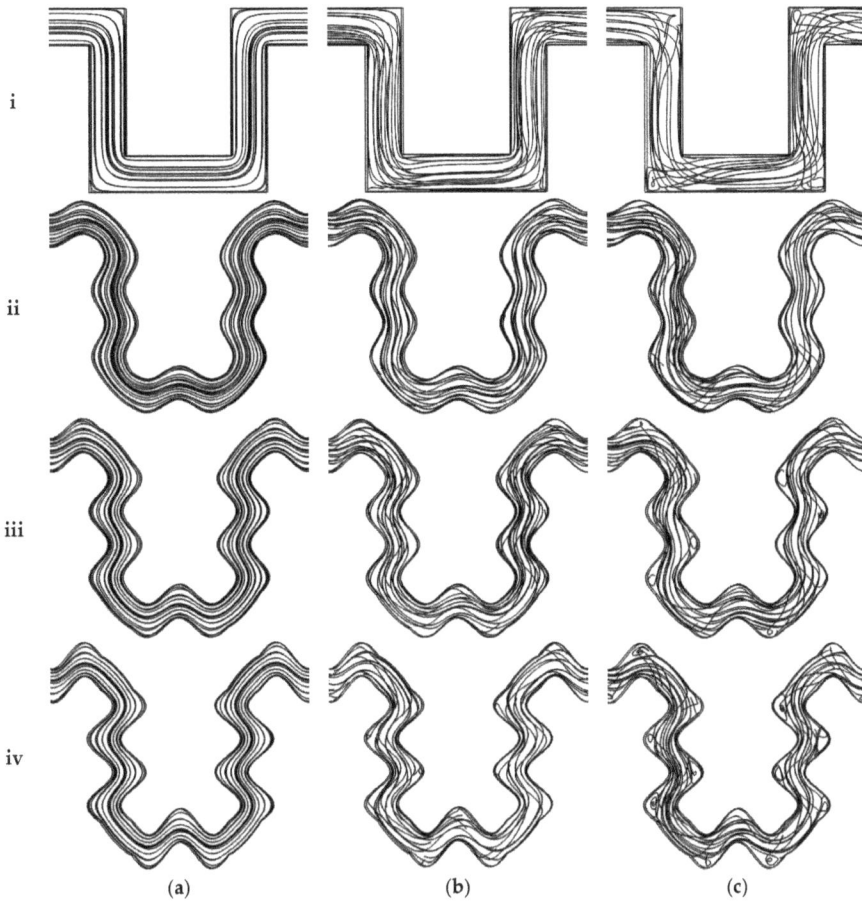

Figure 4. Streamline plots in first U-shaped region of all geometries: (**i**) Case 1; (**ii**) Case 2; (**iii**) Case 3; and (**iv**) Case 4 at (**a**) $Re = 0.1$; (**b**) $Re = 20$; and (**c**) $Re = 50$.

The mixing index variation with the Reynolds number for all geometries is shown in Figure 5. The mixing index is high at Reynolds number of 0.1 despite strictly laminar flow and the absence of streamlines crossing because, at low flow rate, a long residence time of fluids in the channel allows more time for diffusion to occur, which is the dominant factor that causes mixing in this case. At a Reynolds number of 1, the mixing index sharply declines because the residence time decreases with increasing velocity and is insufficient for diffusion. From Reynolds numbers 10 to 30, the mixing index is steady and does not increase with rise in flow rate because secondary flows are in the development stage and are not fully effective in enhancing the mixing of fluids. From Reynolds numbers 1 to 10, the largest increase in the mixing index is observed for sinusoidal walled channels with $A/\lambda = 0.15$ and 0.2 because of secondary flows, that are not fully developed yet, but are still more effective than the other two geometries because of the high amplitude of the sinusoidal walls that would result in a higher

Dean number. The dominant role of secondary flows is evident at Reynolds numbers higher than 30, which increase the mixing index. The mixing index trend with the increase in the Reynolds number for the simple serpentine channel is similar to the previously reported results on simple serpentine geometry [21,29]. Here it is also important to note that the use of sinusoidal walls also increased the total length of the sinusoidal walled serpentine channel compared to the simple serpentine channel. An equal length of both geometries will cause a difference in serpentine shape. Comparison of both geometric shapes with the same total length and the same parameters of serpentine-shaped waves at the same time is not possible. The effect of increased length on mixing performance will be minimal and the enhanced mixing performance of the sinusoidal walled channel is due to the increase in Dean number and secondary flows as explained previously.

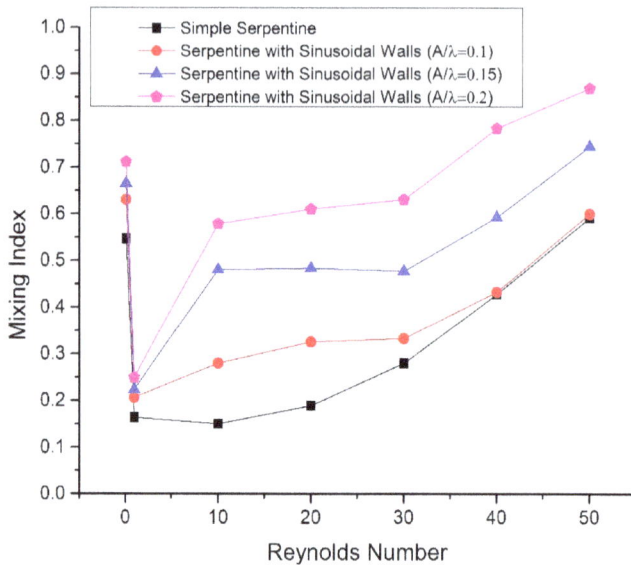

Figure 5. Mixing index vs Reynolds number at the exit.

The concentration distribution of two species at three locations (i.e., planes A–A, B–B, and C–C shown in Figure 1) in all of the geometries at a Reynolds number of 50 is shown in Figure 6. Figure 6a exhibits that both fluid streams are parallel before entering the serpentine region in all geometries with negligible mixing. As fluids move downstream of the channel, secondary flows start developing in the transverse direction, because of turns in the serpentine, and sinusoidal-shaped side walls, thereby enhancing mixing. The concentration distribution becomes more uniform, showing more mixing as the geometric shape changes from simple to sinusoidal walled serpentine and approaches optimal mixing fraction of 0.5 in major portions of the cross section for the maximum amplitude of sinusoidal walls (Figure 6c(iv)). To compare the mixing performance at same total length, the mixing index for fixed stream wise length of 6.2 mm is shown in Figure 7. The mixing trend with same total length for all cases is similar to the mixing trend with the same number of serpentine units but a different total length. The difference occurred at Reynolds numbers after 30 where simple serpentine channel achieved higher mixing index than a sinusoidal walled channel with A/λ ratio of 0.1. This is because the sharp turns of a simple serpentine caused more secondary flow than the curved turns of the sinusoidal walled channel with low amplitude as illustrated in Figure 4c. This effect is dominated by the high amplitude of sinusoids; thus, the two other cases of sinusoidal walled channels demonstrated better performance.

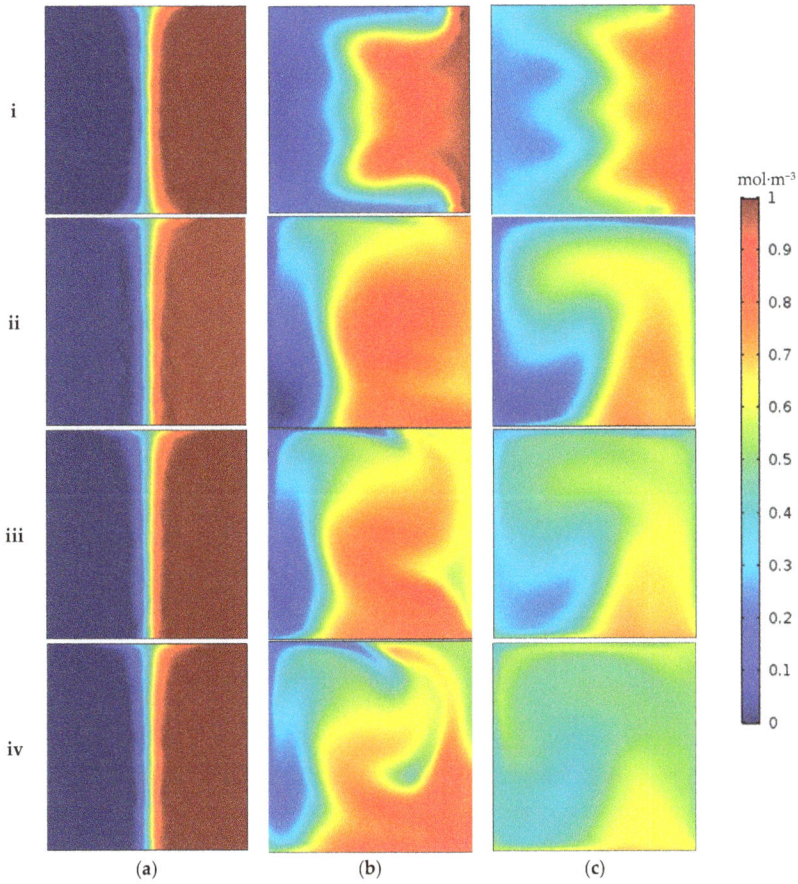

Figure 6. Concentration contours at Reynolds number of 50 at three planes (shown in Figure 1): (**a**) A–A; (**b**) B–B; and (**c**) C–C; for (**i**) Case 1; (**ii**) Case 2; (**iii**) Case 3; and (**iv**) Case 4.

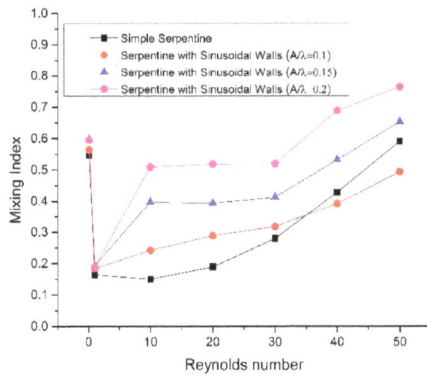

Figure 7. Mixing Index at stream wise length of 6.2 mm.

Velocity arrow plots to show the secondary flow in the transverse direction at Reynolds numbers of 20 and 50 are shown in Figure 8. The cross section showing arrow plots is located at the Plane B–B shown in Figure 1. No secondary flow is observed in the simple serpentine channel at Reynolds number 20; thus, less mixing enhancement is expected, which corroborates the results shown in Figure 5. For the same geometry, two counter rotating vortices appeared in the central region of the cross section at Reynolds number 50, which explain better mixing for this case. For sinusoidal walled serpentine channels, one vortex with its axis near the side wall appears at Reynolds number 20, as shown in Figure 8a. The axis of the vortex moves toward the central region as the amplitude increased. At a Reynolds number of 50, secondary flow is more prominent in the sinusoidal walled channels with two overlapping counter vortical flows, as shown in Figure 8b. The overlapped area spreads across the cross section with the increase in amplitude; thus, a high mixing efficiency of 86.9% is achieved by sinusoidal walled serpentine channel at Reynolds number of 50.

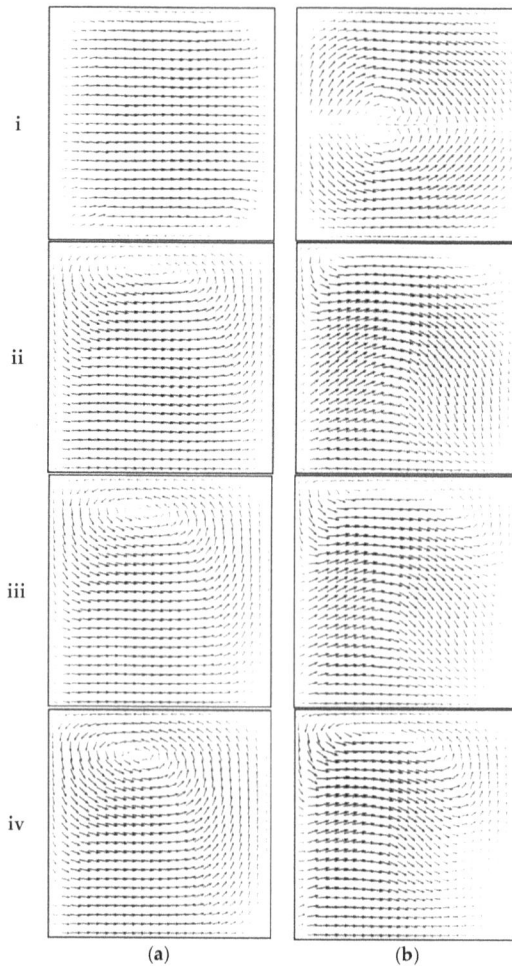

Figure 8. Velocity arrow plots at Plane B-B for all geometries: (**i**) Case 1; (**ii**) Case 2; (**iii**) Case 3; and (**iv**) Case 4 at (**a**) *Re* = 20 and (**b**) *Re* = 50.

Figure 9 shows the pressure drop across all channels, which is proportional to the flow rate and increases with the increase in Reynolds number. Given that the pressure at the outlet is zero, the pressure drop shows the pumping pressure required at the inlet to drive the fluids through the channel. The sinusoidal walled channel with maximum amplitude has the highest pressure drop, followed by the channel with A/λ ratio of 0.15 because sinusoidal walls with high amplitude are more resistant to flow than the other cases. The simple serpentine channel has a slightly higher pressure drop than the sinusoidal walled serpentine channel with A/λ ratio of 0.1 because of a lower resistance to flow from the sinusoidal walls with small amplitude compared to the sharp turns of the simple serpentine channel.

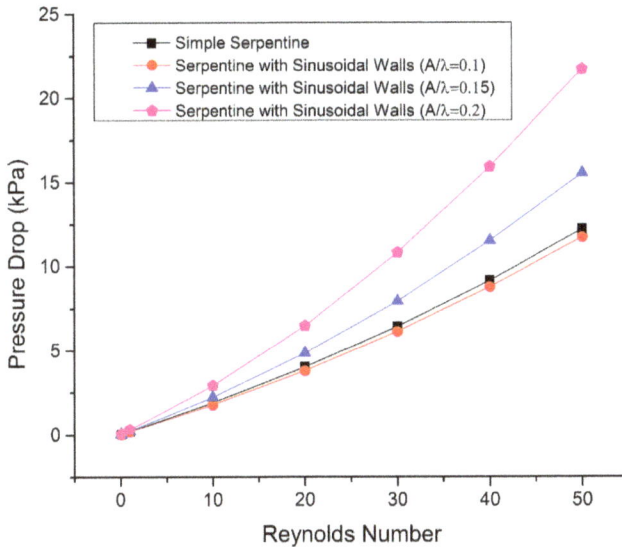

Figure 9. Pressure drop variation with the Reynolds number.

The increase in pumping pressure results in an increased mixing cost because of higher energy consumption. Figure 10a,b show the mixing cost variation with Reynolds number in terms of pressure drop and input power, respectively. A high value of $MI/\Delta P$ ratio represents low mixing cost and better performance. At a Reynolds number of 0.1 mixing cost is low because of diffusion dominated mixing with minimum pressure drop but this is not favorable because of the slow processing time. Rapid mixing can be achieved at high flow rates that cause secondary flows. The increase in flow rate also results in increased pressure drop and thus a rise in energy consumption rate is inevitable. So, for any design, mixing cost will always go up with increased flow rate. High $MI/\Delta P$ ratios of sinusoidal walled channels in Figure 10a show relatively improved mixing performance compared to simple serpentine channels. In the case of mixing cost in terms of input power, a low value represents better mixing performance. The comparison of all cases show that the mixing cost for the three sinusoidal walled cases is similar and comparatively less than the simple serpentine channel except for Reynolds number 50, where geometry with A/λ ratio of 0.2 achieved a high mixing index at a relatively higher mixing cost (Figure 10b). Below a Reynolds number of 40, the mixing cost of sinusoidal walled channels is relatively lower than the simple serpentine channel.

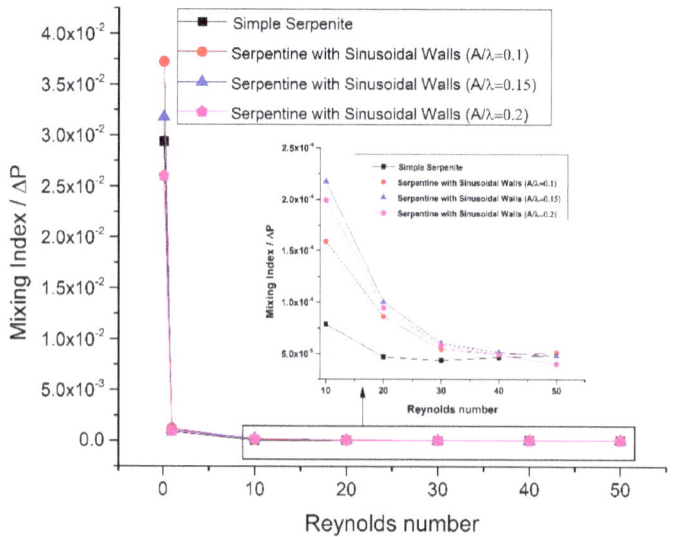

(a)

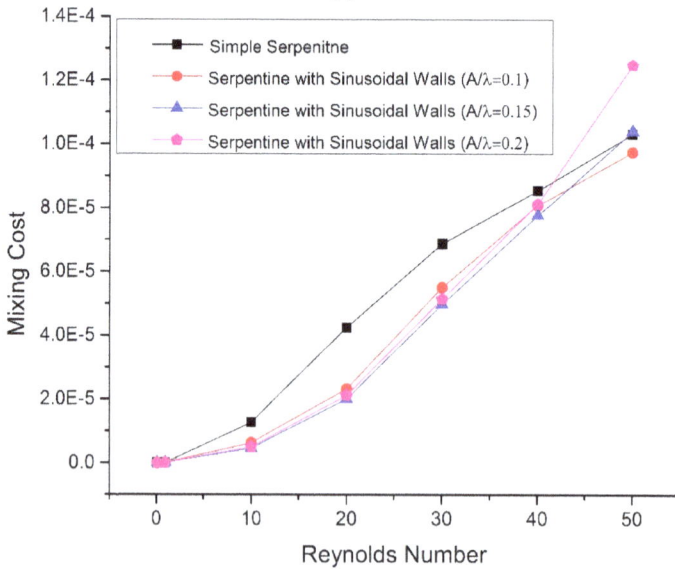

(b)

Figure 10. Mixing cost in terms of (**a**) pressure drop and (**b**) input power.

To characterize rapid mixing, the time required to attain full mixing should also be considered. Although full mixing was not achieved for the number of serpentine waves used in the present study, a comparison based on the mixing index to residence time ratio at different Reynolds numbers is presented in Table 1. A higher mixing index to residence time ratio shows good performance. The residence time was calculated using the length of the serpentine part and average velocity through the channel. Sinusoidal walled channels showed better performance except for a Reynolds number of 40, where simple serpentine showed better performance than sinusoidal walled serpentine with A/λ ratio of 0.1 because of sharp turns.

Micromachines **2018**, *9*, 8

Table 1. Comparison of mixing index to residence time ratio.

Geometry	Mixing Index/Time (s^{-1})		
	$Re = 1$	$Re = 20$	$Re = 40$
Simple serpentine	0.2984	6.87	31.12
Sinusoidal walled ($A/\lambda = 0.1$)	0.3380	10.70	28.43
Sinusoidal walled ($A/\lambda = 0.15$)	0.3395	14.71	36.06
Sinusoidal walled ($A/\lambda = 0.2$)	0.3495	17.11	43.98

7. Conclusions

A chaotic-advection-based serpentine-shaped passive micromixer with sinusoidal side walls was presented in this study, and its mixing performance was evaluated. Numerical simulations were conducted using the Navier–Stokes equations to solve the fluid flow. The resulting velocity field was used to compute the species concentration field using the convection–diffusion equation. The effect of the A/λ ratio of sinusoidal walls on the mixing efficiency of microchannels was also considered. The mixing performance of the proposed sinusoidal walled serpentine design was compared with a simple serpentine channel at Reynolds numbers ranging from 0.1 to 50. At low Reynolds numbers of 0.1 and 1, mixing occurred because of molecular diffusion and depends on the residence time of fluids in the device. Secondary flows started developing with the increase in Reynolds number and showed more effect in the sinusoidal walled serpentine channels. The dominant effect of secondary flow on mixing was observed at Reynolds numbers higher than 30 in all of the geometries. Apart from the increase in flow rate, the increase in A/λ ratio of sinusoidal walls also contributed to the growth of secondary flows because of increase in curvature. The proposed design of serpentine channel with sinusoidal walls achieved a comparatively higher mixing index than the simple serpentine channel with a relatively lower or similar mixing cost.

Acknowledgments: This study was supported by grant from the National Research Foundation of Korea (NRF) funded by the Korean government (MSIP) (No. 2017R1A2B2005515) and a grant from the Priority Research Centers Program through the NRF funded by the MEST (No. 2010-0020089).

Author Contributions: M.U.J. conducted simulations and data analysis and prepared the paper. T.A.C. discussed the results and provided advice. C.W.P. supervised the work and corrected the paper.

Conflicts of Interest: The authors declare no conflict of interest.

References

1. Sackmann, E.K.; Fulton, A.L.; Beebe, D.J. The present and future role of microfluidics in biomedical research. *Nature* **2014**, *507*, 181–189. [CrossRef] [PubMed]
2. Nguyen, N.-T.; Wu, Z. Micromixers—A review. *J. Micromech. Microeng.* **2005**, *15*, R1–R16. [CrossRef]
3. Lee, C.Y.; Chang, C.L.; Wang, Y.N.; Fu, L.M. Microfluidic mixing: A review. *Int. J. Mol. Sci.* **2011**, *12*, 3263–3287. [CrossRef] [PubMed]
4. Ottino, J. Mixing, Chaotic Advection, and Turbulence. *Annu. Rev. Fluid Mech.* **1990**, *22*, 207–253. [CrossRef]
5. Ottino, J.M. *The Kinematics of Mixing: Stretching, Chaos, and Transport*; Cambridge University Press: Cambridge, UK, 1989.
6. Ottino, J.M.; Wiggins, S. Introduction: Mixing in microfluidics. *Philos. Trans. R. Soc. A Math. Phys. Eng. Sci.* **2004**, *362*, 923–935. [CrossRef] [PubMed]
7. Wiggins, S.; Ottino, J.M. Foundations of chaotic mixing. *Philos. Trans. R. Soc. A Math. Phys. Eng. Sci.* **2004**, *362*, 937–970. [CrossRef] [PubMed]
8. Cai, G.; Xue, L.; Zhang, H.; Lin, J. A review on micromixers. *Micromachines* **2017**, *8*, 274. [CrossRef]
9. Abbas, Y.; Miwa, J.; Zengerle, R.; von Stetten, F. Active continuous-flow micromixer using an external braille pin actuator array. *Micromachines* **2013**, *4*, 80–89. [CrossRef]
10. Huang, P.-H.; Xie, Y.; Ahmed, D.; Rufo, J.; Nama, N.; Chen, Y.; Chan, C.Y.; Huang, T.J. An acoustofluidic micromixer based on oscillating sidewall sharp-edges. *Lab Chip* **2013**, *13*, 3847–3852. [CrossRef] [PubMed]

11. Huang, P.-H.; Ren, L.; Nama, N.; Li, S.; Li, P.; Yao, X.; Cuento, R.A.; Wei, C.-H.; Chen, Y.; Xie, Y.; et al. An acoustofluidic sputum liquefier. *Lab Chip* **2015**, *15*, 3125–3131. [CrossRef] [PubMed]

12. Nama, N.; Huang, P.H.; Huang, T.J.; Costanzo, F. Investigation of micromixing by acoustically oscillated sharp-edges. *Biomicrofluidics* **2016**, *10*, 1–17. [CrossRef] [PubMed]

13. Oddy, M.H.; Santiago, J.G.; Mikkelsen, J.C. Electrokinetic instability micromixing. *Anal. Chem.* **2001**, *73*, 5822–5832. [CrossRef] [PubMed]

14. Lee, C.Y.; Wang, W.T.; Liu, C.C.; Fu, L.M. Passive mixers in microfluidic systems: A review. *Chem. Eng. J.* **2016**, *288*, 146–160. [CrossRef]

15. Buchegger, W.; Wagner, C.; Lendl, B.; Kraft, M.; Vellekoop, M.J. A highly uniform lamination micromixer with wedge shaped inlet channels for time resolved infrared spectroscopy. *Microfluid. Nanofluid.* **2011**, *10*, 889–897. [CrossRef]

16. Nimafar, M.; Viktorov, V.; Martinelli, M. Experimental comparative mixing performance of passive micromixers with H-shaped sub-channels. *Chem. Eng. Sci.* **2012**, *76*, 37–44. [CrossRef]

17. Lim, T.W.; Son, Y.; Jeong, Y.J.; Yang, D.-Y.; Kong, H.-J.; Lee, K.-S.; Kim, D.-P. Three-dimensionally crossing manifold micro-mixer for fast mixing in a short channel length. *Lab Chip* **2011**, *11*, 100–103. [CrossRef] [PubMed]

18. SadAbadi, H.; Packirisamy, M.; Wüthrich, R. High performance cascaded PDMS micromixer based on split-and-recombination flows for lab-on-a-chip applications. *RSC Adv.* **2013**, *3*, 7296. [CrossRef]

19. Kim, D.S.; Lee, S.H.; Kwon, T.H.; Ahn, C.H. A serpentine laminating micromixer combining splitting/recombination and advection. *Lab Chip* **2005**, *5*, 739–747. [CrossRef] [PubMed]

20. Mengeaud, V.; Josserand, J.; Girault, H.H. Mixing processes in a zigzag microchannel: Finite element simulations and optical study. *Anal. Chem.* **2002**, *74*, 4279–4286. [CrossRef] [PubMed]

21. Hossain, S.; Ansari, M.A.; Kim, K.Y. Evaluation of the mixing performance of three passive micromixers. *Chem. Eng. J.* **2009**, *150*, 492–501. [CrossRef]

22. Parsa, M.K.; Hormozi, F. Experimental and CFD modeling of fluid mixing in sinusoidal microchannels with different phase shift between side walls. *J. Micromech. Microeng.* **2014**, *24*, 65018. [CrossRef]

23. Parsa, M.K.; Hormozi, F.; Jafari, D. Mixing enhancement in a passive micromixer with convergent-divergent sinusoidal microchannels and different ratio of amplitude to wave length. *Comput. Fluids* **2014**, *105*, 82–90. [CrossRef]

24. Afzal, A.; Kim, K.Y. Convergent-divergent micromixer coupled with pulsatile flow. *Sens. Actuator B-Chem.* **2015**, *211*, 198–205. [CrossRef]

25. Akgönül, S.; Özbey, A.; Karimzadehkhouei, M.; Gozuacik, D.; Koşar, A. The effect of asymmetry on micromixing in curvilinear microchannels. *Microfluid. Nanofluid.* **2017**, *21*, 1–15. [CrossRef]

26. Fan, L.L.; Zhu, X.L.; Zhao, H.; Zhe, J.; Zhao, L. Rapid microfluidic mixer utilizing sharp corner structures. *Microfluid. Nanofluid.* **2017**, *21*, 1–12. [CrossRef]

27. Alam, A.; Afzal, A.; Kim, K.Y. Mixing performance of a planar micromixer with circular obstructions in a curved microchannel. *Chem. Eng. Res. Des.* **2014**, *92*, 423–434. [CrossRef]

28. Liu, R.H.; Stremler, M.A.; Sharp, K.V.; Olsen, M.G.; Santiago, J.G.; Adrian, R.J.; Aref, H.; Beebe, D.J. Passive mixing in a three-dimensional serpentine microchannel. *J. Microelectromech. Syst.* **2000**, *9*, 190–197. [CrossRef]

29. Hossain, S.; Kim, K.Y. Mixing performance of a serpentine micromixer with non-aligned inputs. *Micromachines* **2015**, *6*, 842–854. [CrossRef]

30. Alam, A.; Kim, K.Y. Analysis of mixing in a curved microchannel with rectangular grooves. *Chem. Eng. J.* **2012**, *181–182*, 708–716. [CrossRef]

31. Ottino, J.M. Lamellar mixing models for structured chemical reactions and their relationship to statistical models; Macro- and micromixing and the problem of averages. *Chem. Eng. Sci.* **1980**, *35*, 1377–1381. [CrossRef]

32. Hashmi, A.; Xu, J. On the Quantification of Mixing in Microfluidics. *J. Lab. Autom.* **2014**, *19*, 488–491. [CrossRef] [PubMed]

33. Falk, L.; Commenge, J.M. Performance comparison of micromixers. *Chem. Eng. Sci.* **2010**, *65*, 405–411. [CrossRef]

34. Chung, C.K.; Shih, T.R. A rhombic micromixer with asymmetrical flow for enhancing mixing. *J. Micromech. Microeng.* **2007**, *17*, 2495–2504. [CrossRef]

35. Ortega-Casanova, J. Enhancing mixing at a very low Reynolds number by a heaving square cylinder. *J. Fluids Struct.* **2016**, *65*, 1–20. [CrossRef]

36. Hardt, S.; Schönfeld, F. Laminar mixing in different interdigital micromixers: II. Numerical simulations. *AIChE J.* **2003**, *49*, 578–584. [CrossRef]

micromachines

MDPI

Article

Mixing Enhancement in Serpentine Micromixers with a Non-Rectangular Cross-Section

Joshua Clark, Miron Kaufman and Petru S. Fodor *

Department of Physics, Cleveland state University, 2121 Euclid Avenue, Cleveland, OH 44236, USA;
j.a.clark17@vikes.csuohio.edu (J.C.); m.kaufman@csuohio.edu (M.K.)
* Correspondence: p.fodor@csuohio.edu; Tel.: +1-216-523-7520

Received: 31 January 2018; Accepted: 28 February 2018; Published: 2 March 2018

Abstract: In this numerical study, a new type of serpentine micromixer involving mixing units with a non-rectangular cross-section is investigated. Similar to other serpentine/spiral shaped micromixers, the design exploits the formation of transversal vortices (Dean flows) in pressure-driven systems, associated with the centrifugal forces experienced by the fluid as it is confined to move along curved geometries. In contrast with other previous designs, though, the use of non-rectangular cross-sections that change orientation between mixing units is exploited to control the center of rotation of the transversal flows formed. The associated extensional flows that thus develop between the mixing segments complement the existent rotational flows, leading to a more complex fluid motion. The fluid flow characteristics and associated mixing are determined numerically from computational solutions to Navier–Stokes equations and the concentration-diffusion equation. It is found that the performance of the investigated mixers exceeds that of simple serpentine channels with a more consistent behavior at low and high Reynolds numbers. An analysis of the mixing quality using an entropic mixing index indicates that maximum mixing can be achieved at Reynolds numbers as small as 20 in less than four serpentine mixing units.

Keywords: passive micromixers; Dean flows; serpentine-shaped channels; mixing index

1. Introduction

The use of microfluidic devices in applications ranging from chemical analysis and reaction engineering to biological assays and bioengineering has progressed dramatically in recent years [1–3]. This progress has been fueled by the perceived benefits of employing microfluidic devices; such benefits include reduced reactant consumption, superior heat and mass transfer efficiency enabling increased flexibility in reactor or assay design, field deployability, and scalability [1,4–6]. Their potential for parallel processing has led to new applications in molecular diagnosis [7] and single cell biology [8–12]. Together with the availability of new methodologies for device fabrication, ranging from soft-lithography [13] to 3D printing [14], laser-assisted chemical etching [15], and even paper-based materials [16], this has encouraged an increasing number of researchers to explore this platform and seek new applications.

One of the fundamental operations that microfluidic devices have to achieve as part of their functionality is mixing. Virtually all their applications, including biological/chemical assays, as well as chemical and particulate analysis, require the mixing of two or more component pairs such as analyte/assay or chemical reactants [17]. Since microfluidic devices operate in the low Reynolds number regime, the typical flow characteristics are laminar, with turbulence being absent; thus, the mixing has to rely on diffusional transport. However, this is too slow for many of the envisioned practical applications. The challenge of achieving efficient mixing in microfluidic devices has spurred a large body of research focused on the theoretical, implementation, and fabrication aspects associated with mixing on the microscale. Nguyen [18], Nguyen and Wu [19], Cai et al. [20], and Lee et al. [17] have provided comprehensive reviews of the various strategies employed to address the mixing bottleneck in the development of microfluidic platforms. In brief, micromixers are generally classified as active or passive. The active micromixers use external energy sources such as ultrasonic [21] or acoustic [22–26] vibration, electric fields [27], magnetic stirrers [28], or mechanical actuators [29] in order to stir the fluids of interest. Configurations such as acoustic-based micromixers have been successfully used to achieve rapid mixing, even when highly viscous solutions are involved [25]. Passive micromixers, on the other hand, use only the interaction between the fluid flow and geometrical structures to sequentially laminate and braid the fluids to be mixed or generate cross-sectional mass transport. The first approach relies on increasing the area of contact between the different fluid components and thus on increasing the efficiency of the molecular diffusion mechanism for mixing [30]. The second approach uses an array of geometrical features, such as ridge/groove systems [31–33], obstacles [34], barriers [35], and 2D [36] or 3D [37] turns, to induce transversal advection. It has been shown that, in this case, if the advection induced is chaotic, very fast intermixing between different components can be achieved [18].

While active micromixers can achieve high mixing efficiencies and mixing control over a broad range of Reynolds numbers, they are harder to fabricate, are more difficult to integrate with other microfluidic components, and more importantly require external power sources [20]. Even though in some cases employing complex 3D geometrical structures for mixing control can pose fabrication challenges, for the most part passive micromixers do not suffer from the drawbacks of the active micromixers mentioned above. The absence of an external energy source, aside from the pressure-driven flow, makes them easier to integrate within complex microfluidic systems with standard fabrication methodologies. Moreover, the absence of complex multi-physics interactions that need to be accounted for makes them much more amenable to theoretical or computational modeling. This allows for a more straightforward and efficient optimization process of the various geometrical and flow parameters needed to maximize mixing within various designs [32,38].

A popular design strategy used in passive micromixers to generate cross-sectional flows and induce chaotic advection capable of enhancing the mixing of fluid components, has been the use of channels with repeating curved sections or turns [18]. These systems exploit the centrifugal forces experienced by the fluid as it is guided by the geometry of the channels to move along a curved trajectory. An analysis, first performed by Dean [39], has shown that the flow field that develops inside such a system is consistent with the formation of transversal vortices, also known as Dean flows. These provide a geometrically simple way of promoting advective transport in microchannels using serpentine or spiral-shaped designs. Beside the easy implementation of these designs, additional advantages include the absence of high local shears compared to obstacle-based micromixers as well as the lack of complex 3D surface structures. The first makes them attractive for biological applications targeted at handling without damage to large biomolecules [40]. The second one allows the potential reuse of the devices, as it enables easier cleaning [41]. It has to be noted though that rapid mixing in simple serpentine/spiral designs is achieved typically at large fluid speeds associated with the formation of secondary Dean vortices, leading to the transition to a chaotic advection regime [18]. This corresponds though to Reynolds numbers typically too high for practical settings [42]. In order to achieve more efficient mixing at intermediate and low Reynolds numbers in serpentine micromixers, the effect of various geometrical modifications has been explored. These include the use of grooves on the side or bottom walls of the channel to produce more complex transversal flows [43,44], modulation of the surface of the side walls [45], misaligned inlets for the fluid components to be mixed [46], and three-dimensional turns [47].

In this computational work, we report on the effect on mixing when non-rectangular cross-sections are used for the curved sections forming a simple serpentine channel. The use of this type of section as its orientation along the channel is changed allows for the centers of rotation of the primary Dean vortexes that form to be shifted between the mixing units. The effect is akin to that achieved in Strook micromixers [31,48], where the use on the floor of the straight channels of asymmetric ridges with variable apex positions is exploited to achieve extensional flows and induce chaotic advection.

From an implementation point of view, the designs proposed retain the benefits associated with serpentine micromixers such as simple fabrication, easy cleaning, and potentially damage-free processing of biological samples containing large molecules. The numerical work discussed is performed for a range of Reynolds numbers from $Re = 1$ to 100 for both the new design and the design of the standard serpentine micromixer.

2. Geometrical Design of the Micromixer

The basic designs of serpentine micromixers used in this computational study are shown in Figures 1 and 2. The fluids to be mixed are fed into the mixer through a T-shaped inlet structure. The shape of both inlets is square with a size of 50 μm. For both designs, each mixing unit consists of two semicircular sections connected by a straight one. For all the channels investigated, the total height of the main channels is $H = 100$ μm, while its total width is $W = 200$ μm. The length of the straight sections connecting subsequent turns is set to be equal with W. For the non-rectangular cross-section designs, the height of the thinner part is maintained constant at $H/2 = 50$ μm, while its width is set at $W/2 = 100$ μm. As shown in Figure 2, between each turn of the serpentine, the orientation of the non-rectangular cross-section is changed so that the thicker part of the channels is always on the outside of the turns. For this study, the fluid flow and mixing performance have been analyzed for values of the inner turn radius R_{in} corresponding to 0.5 W for both the standard rectangular section micromixer as well as for the newly investigated design.

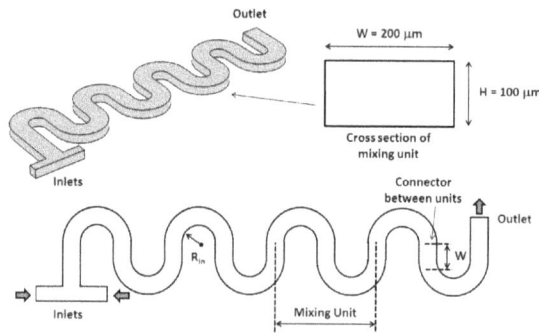

Figure 1. 3D and top views of the standard serpentine micromixer with a rectangular cross-section defined by $W = 200$ μm and $H = 100$ μm.

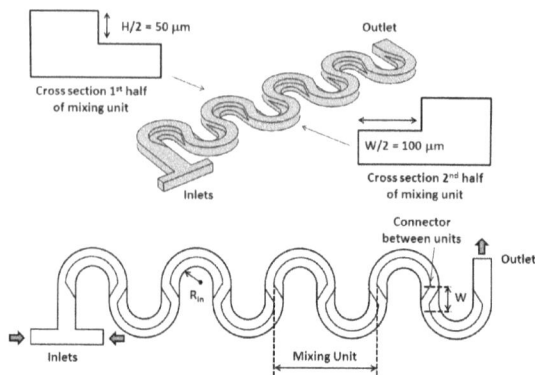

Figure 2. 3D and top views of the new serpentine micromixer design employing non-rectangular cross-sections. As shown above, the orientation of the cross-section of the mixer is changed after each turn of the serpentine.

3. Numerical Model and Mixing Assessment

The flow fields for each channel are obtained by solving the Navier–Stokes equations of motion for an incompressible Newtonian fluid in steady state pressure-driven flow:

$$\rho\left[\frac{\partial u}{\partial t} + (u \cdot \nabla)u\right] = -\nabla p + \eta \nabla^2 u \tag{1}$$

$$\nabla \cdot u = 0 \tag{2}$$

where u (m·s^{-1}) is the velocity vector, ρ (kg·m^3) is the fluid density, η (kg·m^{-1}·s^{-1}) is the fluid viscosity, t (s) is the time, and p (Pa) is the pressure. The flow field equations are solved using a generalized minimal residual method (GMRES) iterative solver with a geometrical multigrid pre-conditioner and a Vanka algorithm for the pre- and post-smoothing. No-slip boundary conditions were set for the walls of the micromixer. A free tetrahedral mesh is used for the entire microchannel with no less than ~200,000 elements for all the geometries studied.

The fluid speeds and pressures thus obtained are then used to compute the species concentration throughout the micromixers using the convection-diffusion equation:

$$\frac{\partial c}{\partial t} = D\nabla^2 c - u \cdot \nabla c \tag{3}$$

where c (mol·m^{-3}) is the concentration of the species of interest, and D (m^2·s^{-1}) is its diffusion constant, respectively. The same iterative numerical solver as for the Navier–Stokes equations is used, but the maximum element size in the mesh is constrained to less than 10 µm through the whole geometry to avoid the possible numerical errors that can be associated with this type of solution. Thus, for all the concentration simulations, the number of mesh elements in the numerical model has never been less than ~1,900,000. For all the simulations described in this work, we used the computational package COMSOL Multiphysics 5.1 (COMSOL Inc., Stockholm, Sweden) and its computational fluid dynamics/chemical engineering module. The accuracy of the numerical work employed for this type of flow has been previously validated against data obtained from other microfluidic devices [49,50]. In particular, when comparing model data with measured data on similar planar curved microchannels, as presented by Jiang et al. [50], we find mixing times to be less than 0.05 s for Dean numbers equal to or above 140, in agreement with the experimental results.

To quantify the mixing quality, we use a mixing index based on calculating the Shannon entropy associated with the distribution of the various components across sections perpendicular to the microchannels. Previous work has shown this type of measure to be consistent with other mixing measures and, at the same time, to be more mathematically rigorous and easily adaptable to the various formats in which computational and experimental data is presented, such as concentration distributions, particle tracer Poincaré maps, or fluorescent intensity images [51,52]. In all of these cases, the data of interest is converted to image data, and the informational complexity of these images is then analyzed to quantify the level of segregation or mixing of the components of interest.

In brief, to quantify the mixing, the obtained cross-sectional concentration distributions at different positions along the channel are first converted to 8-bit grayscale intensity maps [53]. These images are divided into a number N_{bins} of equal size regions. For a system with two components, the mixing index is defined as

$$M = -\frac{1}{ln2} \cdot \frac{1}{N_{bins}} \cdot \sum_{j=1}^{N_{bins}} \left[p_{1/j}ln\left(p_{1/j}\right) + p_{2/j}ln\left(p_{2/j}\right)\right] \tag{4}$$

where $p_{1/j}$ and $p_{2/j}$ are the conditional probabilities for Components 1 and 2, respectively, to be located in bin j. They represent the fraction of Components 1 and 2, respectively, in each bin relative to the total. They are calculated as the ratio of the average bin grayscale intensity from the corresponding concentration image, normalized by the maximum intensity, i.e., 255 for grayscale image data. In this particular study, since we have two chemical species and the fluids used are incompressible the two conditional probabilities are related as $p_{2/j} = 1 - p_{1/j}$. As shown in Equation (4), the mixing index M is normalized by a factor of $ln2$, where 2 corresponds to the number of components. Thus, the mixing index will take the value $M = 0$ for completely segregated components, while it will assume the value $M = 1$ for the completely mixed case.

4. Results and Discussion

For the analysis of all the mixers, water solutions have been considered as the working fluids, each with a density ρ of 1000 kg·m^{-3} and a viscosity η of 0.001 kg·m^{-1}·s^{-1}. The diffusion coefficient D was fixed to 1.0×10^{-9} m^2·s^{-1} corresponding to the diffusion values for most ions in aqueous solutions. For all the simulations, the working solutions were considered to be pure water introduced through one of the inlets and dyed water with a concentration $c = 1$ moL·m^{-3} introduced through the opposite inlet. The flow rates are maintained for both fluid components, with the same mean fluid velocity at both inlets. This means fluid velocity is varied from 0.0075 to 0.75 m·s^{-1} to span a range of Reynolds numbers corresponding to $Re = 1$–100.

Representative results for the flow fields and concentration distributions for the designs investigated are shown in Figures 3 and 4. From the evolution of the concentration along the length of the micromixer, it is immediately apparent that its distribution is distinct between the two designs.

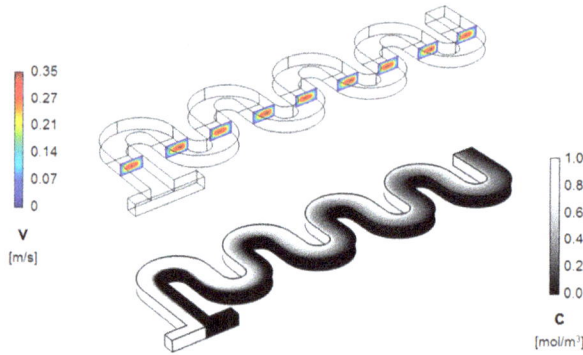

Figure 3. Velocity magnitude (cross-sectional maps) (**top**) and concentration distribution (surface map) (**bottom**) along the channel of a standard serpentine micromixer ($R_{in} = W/2 = 100$ μm, and $Re = 20$).

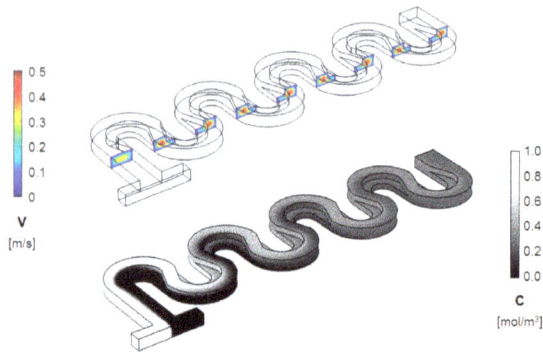

Figure 4. Velocity magnitude (cross-sectional maps) (**top**) and concentration distribution (surface map) (**bottom**) along the channel of a non-rectangular cross-section serpentine micromixer ($R_{in} = W/2 = 100$ μm, and $Re = 20$).

As shown in the figures above, in the standard design, the two fluids, which are introduced in the system through the opposite ends of the T-joint inlet, remain on distinctive sides of the serpentine through the length of the device, for Reynolds numbers as high as $Re = 30$. On the other hand, for the serpentine with non-rectangular cross-sections, that change orientation between the serpentine sections, the concentration distribution starts to homogenize after the first mixing unit. This conclusion from the concentration surface maps is supported by the cross-sectional concentration maps. As shown in Figures 5 and 6, the concentration distribution transversal to the flow illustrate different behaviors in the two designs. While, in both microchannel types, fluid motion does occur transversally as expected under the presence of Dean flows in curved geometries [18], in the regular serpentine channel, the two fluids remain mostly separated. The interface between the fluid streams does deform, which increases the contact area between them, but the mixing remains limited to the molecular diffusion at the boundary. In contrast to the modified design, aside from the interface stretching, sizable segmentation of the fluid streams also occurs, at Reynolds numbers as low as $Re = 20$. In fact, based on the evolution of the concentration along the channel (Figures 5 and 6), as soon as the third mixing cycle occurs, there are virtually no pockets of low ($c = 0$ mol·m^{-3}) and high ($c = 1$ moL·m^{-3}) concentration, respectively, left in the dyed fluid distribution map.

This visual assessment of the concentration distribution in the two types of micromixers is also confirmed by the quantitative analysis of the mixing achieved in them. Graphs of the mixing index (Figure 7) indicated that, as the fluids flow along the channel, their intermixing increases. Additionally, increasing the fluid rates (Reynolds number) is associated with better mixing performance. The mixing increase with Re is consistent with previous analyses of Dean flow micromixers [18,52], which showed that the Dean vortices that formed in curved microchannels become increasingly complex and undergo bifurcations at high Reynolds numbers, which are associated with the onset of chaotic advection. However, while, in the regular serpentine micromixer, full mixing ($M = 1$) is not achieved except in the high Reynolds number regime ($Re = 100$), the modified designs show very good mixing behavior across the full range of inlet fluid rates investigated. In the non-rectangular cross-section mixers, full mixing is realized within three mixing units down to Reynolds numbers $Re = 20$. This corresponds to a mixing time of less than 0.035 s for Reynolds numbers larger than 20. Moreover, even at Reynolds numbers as low as 1, no mixing saturation is observed, indicating that full mixing is achievable by adding more mixing units to the channel.

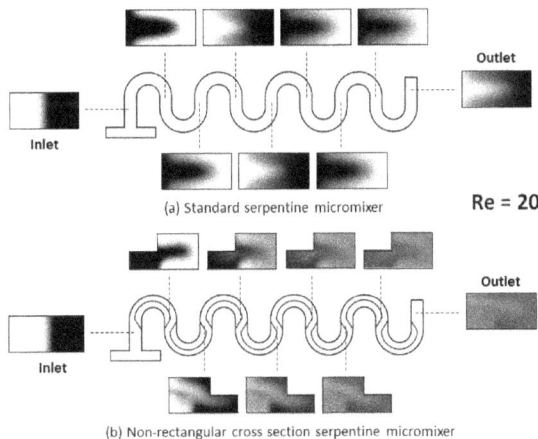

Figure 5. Transversal concentration distribution at various positions along the channel for (**a**) a standard serpentine micromixer and (**b**) a serpentine micromixer with a non-rectangular cross-section ($Re = 20$). The concentration is mapped at both the midpoint and the end of each mixing unit.

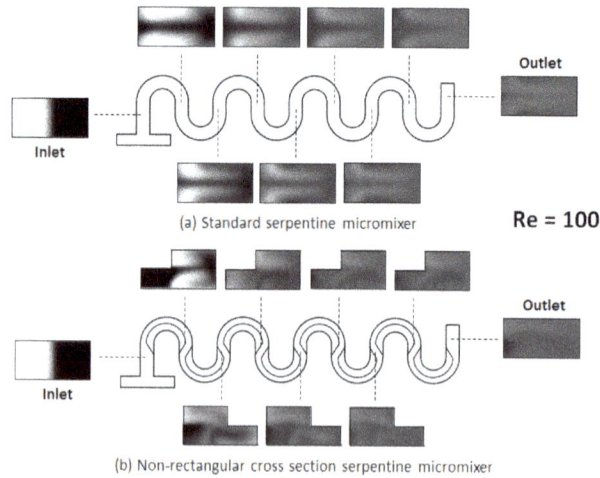

Figure 6. Transversal concentration distribution at various positions along the channel for (**a**) a standard serpentine micromixer and (**b**) a serpentine micromixer with a non-rectangular cross-section (*Re* = 100).

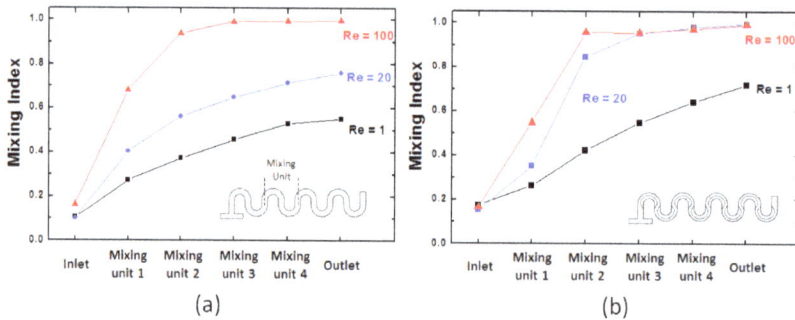

Figure 7. Position dependence of the mixing index along serpentine micromixers at different Reynolds numbers: (**a**) a standard serpentine microchannel and (**b**) a modified channel.

From the mixing quality dependence on the position along the channel, it also has to be noted that, while in the standard serpentine channel, the rate of increase in the mixing index is almost monotonic as a function of the length of the channel, this is not the case for the non-rectangular section designs. Essentially after the first mixing unit for Reynolds numbers larger than *Re* = 20, there is a steep increase in mixing quality. This behavior is similar to that encountered in grooved chaotic advection micromixers [31], where changing the center of the transversal rotation of the fluids between mixing units leads to extensional flows that increase both the inter-lamination of the components to be mixed and their cross-flow. Previous work on this effect in mixers systems with asymmetric slanted groove (also known as staggered herringbone micromixers) [33] has found similar steep power law type increases in mixing performance, indicative of the onset of chaotic advection. While the mechanism for generating pressure-driven transversal flows is different, i.e., using slanted grooves with respect to the flow direction, versus constraining the fluid to move along curved channels, the idea of the approach used remains the same.

Arrow plots based on the velocity fields in the serpentine-based designs confirm that the confinement of trajectories along the curved lines induces counter rotating transversal flows. As shown in Figure 8, for the non-rectangular cross-section microchannels, the topology used allows for the center of rotation of these vortices to be changed between each half mixing unit. More importantly, the topology used leads to the generation of secondary vortexes once the second mixing cycle begins. While secondary vortexes can be created in regular curved Dean micromixers, their onset is associated with large Reynolds numbers, typically $Re > 150$ [31]. Our analysis indicates that the use of the proposed geometry allows the observed bifurcation of multiple vortexes to occur at much lower Reynolds numbers, providing a method of increasing the mixing performance of this type of device that is easy to implement. The proposed devices can be fabricated using soft-lithography techniques [13,54] based on replica molding transfer to polydimethylsiloxane (PDMS) from silicon stamps prepared using two-layer lithography. Due to the smaller non-rectangular cross-section of the channels, the pressure gradient in these devices will be a factor of ~1.9 larger than in their standard counterparts; nevertheless, this remains within the working range of PDMS-based devices.

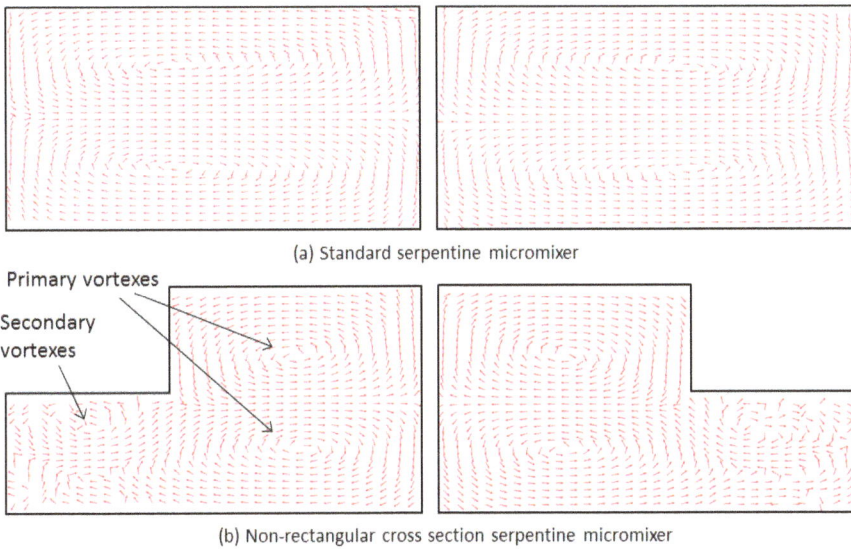

(a) Standard serpentine micromixer

(b) Non-rectangular cross section serpentine micromixer

Figure 8. Arrow plots of the velocity field for the two mixers design: (**a**) a standard micromixer and (**b**) a non-rectangular cross-section micromixer. Both sets of data are collected for the 2nd mixing cycle at $Re = 20$. The left plots are for data collected at the midpoint of the first leg of the mixing cycle, while the right images are the corresponding plots for the second leg of the mixing cycle.

5. Conclusions

In this work, we have investigated the use of a simple adaptation of serpentine-type micromixers in which channels with non-rectangular cross-sections are used. Numerical analysis of the quality of mixing that can be achieved in these micromixers indicates that the use of this topology, coupled with the change in orientation of the channel cross-section between mixing units, increased the mixing performance in terms of a low and intermediate Reynolds numbers for micromixers based on Dean flows. The simplicity of the modification proposed makes it easy to integrate typical serpentine designs used in microfluidic devices, whose applicability was limited by the large Reynolds numbers needed for efficient mixing. Future work will focus on optimization in terms of mixing efficiency of the geometrical parameters used (the size of the non-rectangular cross-section, the radius of curvature of the mixing units, and the length of the connector between them) as well as on the inlet conditions (the ratio of the flows and the nature of fluid components entering the micromixer).

Acknowledgments: This study was supported with funding from the Cleveland State University Undergraduate Summer Research Award 2017, Undergraduate Research Award 2018, and Graduate Faculty Research Award 2018 programs.

Author Contributions: Joshua Clark conducted simulations and data analysis. Miron Kaufman contributed to data analysis using an entropic index, discussed the results, and corrected the paper. Petru S. Fodor developed and supervised the study, conducted data analysis, and prepared the paper.

Conflicts of Interest: The authors declare no conflict of interest.

References

1. Sackmann, E.K.; Fulton, A.L.; Beebe, D.J. The present and future role of microfluidics in biomedical research. *Nature* **2014**, *507*, 181–189. [CrossRef] [PubMed]
2. Capretto, L.; Carugo, D.; Mazzitelli, S.; Nastruzzi, C.; Xunli, Z. Microfluidic and lab-on-a-chip preparation routes for organic nanoparticles and vesicular systems for nanomedicine applications. *Adv. Drug Deliv. Rev.* **2013**, *65*, 1496–1543. [CrossRef] [PubMed]
3. Chin, C.D.; Linder, V.; Sia, S.K. Lab-on-a-chip devices for global health: Past studies and future opportunities. *Lab Chip* **2007**, *7*, 41–57. [CrossRef] [PubMed]
4. Fair, R.B. Digital microfluidics: Is a true lab-on-a-chip possible? *Microfluid. Nanofluid.* **2007**, *3*, 245–281. [CrossRef]
5. Geong, G.S.; Chung, S.; Kim, C.B. Applications of micromixing technology. *Analyst* **2010**, *135*, 460–473. [CrossRef]
6. Chiu, D.T.; deMello, A.J.; Di Carlo, D.; Doyle, P.S.; Hansen, C.; Maceiczyk, R.M.; Wootton, R.C.R. Small but perfectly formed? Successes, challenges, and opportunities for microfluidics in the chemical and biological sciences. *Chem* **2017**, *2*, 201–223. [CrossRef]
7. Jayamohan, H.; Sant, H.J.; Gale, B.K. Applications of microfluidics for molecular diagnostics. *Methods Mol. Biol.* **2013**, *949*, 305–334. [CrossRef] [PubMed]
8. Du, W.; Li, L.; Nichols, K.P.; Ismagilov, R.F. SlipChip. *Lab Chip* **2009**, *9*, 2286–2292. [CrossRef] [PubMed]
9. Bamford, R.A.; Smith, A.; Metz, J.; Glover, G.; Titball, R.W.; Pagliara, S. Investigating the physiology of viable but non-culturable bacteria by microfluidics and time-lapse microscopy. *BMC Biol.* **2017**, *15*, 121. [CrossRef] [PubMed]
10. Zilionis, R.; Nainys, J.; Veres, A.; Savova, V.; Zemmour, D.; Klein, A.M.; Mazutis, L. Single-cell barcoding and sequencing using droplet microfluidics. *Nat. Protoc.* **2017**, *12*, 44–73. [CrossRef] [PubMed]
11. Junkin, M.; Tay, S. Microfluidic single-cell analysis for systems immunology. *Lab Chip* **2014**, *14*, 1246–1260. [CrossRef] [PubMed]
12. Pagliara, S.; Franze, K.; McClain, C.R.; Wylde, G.W.; Fisher, C.L.; Franklin, R.J.M.; Kabla, A.J.; Keyser, U.F.; Chalut, K.J. Auxetic nuclei in embryonic stem cells exiting pluripotency. *Nat. Mater.* **2014**, *13*, 638–644. [CrossRef] [PubMed]
13. Tsao, C.-W. Polymer microfluidics: Simple, low-cost fabrication process bridging academic lab research to commercialized production. *Micromachines* **2016**, *7*, 225. [CrossRef]
14. Waheed, S.; Cabot, J.M.; Macdonald, N.P.; Lewis, T.; Guijt, R.M.; Paullab, B.; Breadmore, M.C. 3D printed microfluidic devices: Enablers and barriers. *Lab Chip* **2016**, *11*, 1993–2013. [CrossRef] [PubMed]
15. Shan, C.; Chen, F.; Yang, Q.; Jiang, Z.; Hou, X. 3D multi-microchannel helical mixer fabricated by femtosecond laser inside fused silica. *Micromachines* **2018**, *9*, 29. [CrossRef]
16. Sun, G.; Panpan, W.; Shenguang, G.; Lei, G.; Jinghua, Y.; Mei, Y. Photoelectrochemical sensor for pentachlorophenol on microfluidic paper-based analytical devicebased on the molecular imprinting technique. *Biosens. Bioelectron.* **2014**, *56*, 97–103. [CrossRef] [PubMed]
17. Lee, C.Y.; Chang, C.L.; Wang, Y.N.; Fu, L.M. Microfluidic mixing: A review. *Int. J. Mol. Sci.* **2011**, *12*, 3263–3287. [CrossRef] [PubMed]
18. Nguyen, N.-T. *Micromixers: Fundamentals, Design and Fabrication*, 2nd ed.; Elsevier: Oxford, UK, 2012; ISBN 978-1-43-773520-8.
19. Nguyen, N.-T.; Wu, Z. Mixers—A review. *J. Micromech. Microeng.* **2005**, *15*, R1–R16. [CrossRef]
20. Cai, G.; Xue, L.; Zhang, H.; Lin, J. A review of micromixers. *Micromachines* **2018**, *8*, 274. [CrossRef]

21. Brandhoff, L.; Zirath, H.; Salas, M.; Haller, A.; Peham, J.; Wiesinger-Mayr, H.; Spittler, A.; Schnetz, G.; Lang, W.; Vellekoop, M.J. A multi-purpose ultrasonic streaming mixer for integrated magnetic bead ELISAs. *J. Micromech. Microeng.* **2015**, *25*, 104001. [CrossRef]

22. Phan, H.V.; Coskun, M.B.; Sesen, M.; Pandraud, G.; Neild, A.; Alan, T. Vibrating membrane with discontinuities for rapid and efficient microfluidic mixing. *Lab Chip* **2015**, *15*, 4206–4216. [CrossRef] [PubMed]

23. Nama, N.; Huang, P.-H.; Huang, T.J.; Constanzo, F. Investigation of micromixing by acoustically oscillated sharp-edges. *Biomicrofluidics* **2016**, *10*, 024124. [CrossRef] [PubMed]

24. Patel, M.V.; Tovar, A.R.; Lee, A.P. Lateral cavity acoustic transducer as an on-chip cell/particle microfluidic switch. *Lab Chip* **2012**, *12*, 139–145. [CrossRef] [PubMed]

25. Huang, P.-H.; Ren, L.; Nama, N.; Li, S.; Li, P.; Yao, X.; Cuento, R.A.; Wei, C.-H.; Chen, Y.; Xie, Y.; et al. An acoustofluidic sputum liquefier. *Lab Chip* **2015**, *15*, 3125–3131. [CrossRef] [PubMed]

26. Destgeer, G.; Im, S.; Ha, B.H.; Jung, J.H.; Ansari, M.A.; Sung, H.J. Adjustable, rapidly switching microfluidic gradient generation using focused travelling surface acoustic wave. *Appl. Phys. Lett.* **2014**, *104*, 023506. [CrossRef]

27. Krishnaveni, T.; Renganathan, T.; Picardo, J.R.; Pushpavanam, S. Numerical study of enhanced mixing in pressure-driven flows in microchannels using a spatially periodic electric field. *Phys. Rev. E* **2017**, *96*, 033117. [CrossRef] [PubMed]

28. Ryu, K.S.; Shaikh, K.; Goluch, E.; Fana, Z.; Liu, C. Micro magnetic stir-bar mixer integrated with parylene microfluidic channels. *Lab Chip* **2004**, *6*, 608–613. [CrossRef] [PubMed]

29. Abbas, Y.; Miwa, J.; Zengerle, R.; von Stetten, F. Active continuous-flow micromixer using an external braille pin actuator array. *Micromachines* **2013**, *4*, 80–89. [CrossRef]

30. Tofteberg, T.; Skolimowski, M.; Andreassen, E.; Geschke, O. A novel passive micromixer: Lamination in a planar channel system. *Microfluid. Nanofluid.* **2010**, *8*, 209–215. [CrossRef]

31. Stroock, A.D.; Dertinger, S.K.W.; Ajdari, A.; Mezic, I.; Stone, H.A.; Whitesides, G.M. Chaotic mixer for microchannels. *Science* **2002**, *295*, 647–651. [CrossRef] [PubMed]

32. Kee, S.P.; Gavriilidis, A. Design and characterization of the staggered herringbone mixer. *Chem. Eng. J.* **2008**, *142*, 109–121. [CrossRef]

33. Fodor, P.S.; Kaufman, M. The evolution of mixing in the staggered herring bone micromixer. *Mod. Phys. Lett. B* **2011**, *25*, 1111–1125. [CrossRef]

34. Alam, A.; Afzal, A.; Kim, K.-Y. Mixing performance of a planar micromixer with circular obstructions in a curved microchannel. *Chem. Eng. Res. Des.* **2014**, *92*, 423–434. [CrossRef]

35. Kim, D.S.; Lee, S.W.; Kwon, T.H.; Lee, S.S. A barrier embedded chaotic micromixer. *J. Micromech. Microeng.* **2004**, *15*, 798–805. [CrossRef]

36. Scherr, T.; Quitadamo, C.; Tesvich, P.; Park, D.S.; Tiersch, T.; Hayes, D.; Choi, J.W.; Nandakumar, K.; Monroe, W.T. A planar microfluidic mixer based on logarithmic spirals. *J. Micromech. Microeng.* **2012**, *22*, 055019. [CrossRef] [PubMed]

37. Chen, X.; Li, T. A novel design for passive misscromixers based on topology optimization method. *Biomed. Microdevices* **2016**, *18*, 57. [CrossRef] [PubMed]

38. Shamloo, A.; Madadelahi, M.; Akbari, A. Numerical simulation of centrifugal serpentine micromixers and analyzing mixing quality parameters. *Chem. Eng. Process. Process Intensif.* **2016**, *104*, 243–252. [CrossRef]

39. Dean, W.R. Note on the motion of a fluid in a curved pipe. *Philos. Mag.* **1927**, *4*, 208–223. [CrossRef]

40. Liu, R.H.; Stremler, M.A.; Sharp, K.V.; Olsen, M.G.; Santiago, J.G.; Adrian, R.J.; Aref, H.; Beebe, D.J. Passive mixing in a three-dimensional serpentine microchannel. *J. Microelectromech. Syst.* **2000**, *9*, 190–197. [CrossRef]

41. Araci, I.E.; Robles, M.; Quake, S.R. A reusable microfluidic device provides continuous measurement capability and improves the detection limit of digital biology. *Lab Chip* **2016**, *16*, 1573–1578. [CrossRef] [PubMed]

42. Mengeaud, V.; Josserand, J.; Girault, H.H. Mixing processes in a zigzag microchannel: Finite element simulation and optical study. *Anal. Chem.* **2002**, *74*, 4279–4286. [CrossRef] [PubMed]

43. Alam, A.; Kim, K.Y. Analysis of mixing in a curved microchannel with rectangular grooves. *Chem. Eng. J.* **2012**, *181–182*, 708–716. [CrossRef]

44. Cook, K.J.; Fan, Y.; Hassan, I. Mixing evaluation of a passive scaled-up serpentine micromixer with slanted grooves. *J. Fluids Eng.* **2013**, *135*, 081102. [CrossRef]

45. Javaid, M.U.; Cheema, T.A.; Park, C.W. Analysis of passive mixing in a serpentine microchannel with sinusoidal side walls. *Micromachines* **2018**, *9*, 8. [CrossRef]
46. Hossain, S.; Kim, K.-Y. Mixing performance of a serpentine micromixer with non-aligned inputs. *Micromachines* **2015**, *6*, 842–854. [CrossRef]
47. Sayah, A.; Gijs, M.A.M. Understanding the mixing process in 3D microfluidic nozzle/diffuser systems: Simulations and experiments. *J. Micromech. Microeng.* **2016**, *26*, 115017. [CrossRef]
48. Fodor, P.S.; Itomlenskis, M.; Kaufman, M. Assessment of mixing in passive microchannels with fractal surface patterning. *Eur. Phys. J. Appl. Phys.* **2009**, *47*, 31301. [CrossRef]
49. D'Alessandro, J.; Fodor, P.S. Use of grooved microchannels to improve the performance of membrane-less fuel cells. *Fuel Cells* **2014**, *14*, 818–826. [CrossRef]
50. Jiang, F.; Drese, K.S.; Hardt, S.; Küpper, M.; Schönfeld, F. Helical flows and chaotic mixing in curved micro channels. *AIChE J.* **2004**, *50*, 2297–2305. [CrossRef]
51. Camesasca, M.; Kaufman, M.; Manas-Zloczower, I. Quantifying fluid mixing with the Shannon entropy. *Macromol. Theory Simul.* **2006**, *15*, 595–607. [CrossRef]
52. Fodor, P.S.; Vyhnalek, B.; Kaufman, M. Entropic Evaluation of Dean Flow Micromixers. In Proceedings of the 2013 COMSOL Conference, Boston, MA, USA, 9–11 October 2013. Available online: https://www.comsol.com/paper/entropic-evaluation-of-dean-flow-micromixers-15053 (accessed on 21 January 2018).
53. Alemaskin, K.; Manas-Zloczover, I.; Kaufman, M. Entropic analysis of color homogeneity. *Polym. Eng. Sci.* **2005**, *45*, 1031–1038. [CrossRef]
54. Xia, Y.; Whitesides, G.M. Soft lithography. *Annu. Rev. Mater. Sci.* **1998**, *28*, 153–184. [CrossRef]

micromachines

MDPI

Article

Numerical and Experimental Study on Mixing Performances of Simple and Vortex Micro T-Mixers

Mubashshir Ahmad Ansari [1], Kwang-Yong Kim [2,*]and Sun Min Kim [2,3,*]

1   Department of Mechanical Engineering, Zakir Husain College of Engineering and Technology,
    Aligarh Muslim University, Aligarh 202001, India; mub.ansari@yahoo.com
2   Department of Mechanical Engineering, Inha University, Incheon 22212, Korea
3   WCSL of Integrated Human Airway-on-a-Chip, Inha University, Incheon 22212, Korea
*   Correspondence: kykim@inha.ac.kr (K.-Y.K.); sunmk@inha.ac.kr (S.M.K.);
    Tel.: +82-32-872-3096 (K.-Y.K.); +82-32-860-7328 (S.M.K.);
    Fax: +82-32-868-1716 (K.-Y.K.); +82-32-877-7328 (S.M.K.)

Received: 30 March 2018; Accepted: 25 April 2018; Published: 27 April 2018

Abstract: Vortex flow increases the interface area of fluid streams by stretching along with providing continuous stirring action to the fluids in micromixers. In this study, experimental and numerical analyses on a design of micromixer that creates vortex flow were carried out, and the mixing performance was compared with a simple micro T-mixer. In the vortex micro T-mixer, the height of the inlet channels is half of the height of the main mixing channel. The inlet channel connects to the main mixing channel (micromixer) at the one end at an offset position in a fashion that creates vortex flow. In the simple micro T-mixer, the height of the inlet channels is equal to the height of the channel after connection (main mixing channel). Mixing of fluids and flow field have been analyzed for Reynolds numbers in a range from 1–80. The study has been further extended to planar serpentine microchannels, which were combined with a simple and a vortex T-junction, to evaluate and verify their mixing performances. The mixing performance of the vortex T-mixer is higher than the simple T-mixer and significantly increases with the Reynolds number. The design is promising for efficiently increasing mixing simply at the T-junction and can be applied to all micromixers.

Keywords: micromixer; vortex micro T-mixer; planar serpentine microchannel; microfluidics

---

1. Introduction

Recently, microfluidic devices have gained popularity for the development of miniaturized analysis systems with wider applications like chemical, biochemical reactions, biomedical devices and drug delivery [1–13]. Manipulating microparticles/cells and small volume of fluids has lead us to the development of microdevices with specific functionalities, like separation/sorting and mixing of fluids. Testing or analysis of microfluidics systems can be performed with reduced cost and time. Rapid mixing of fluids is an essential process for the homogenization of the samples in these microfluidics devices. Recently, numerous research works have been carried out on the design and development of new micromixers. Micromixers can be broadly categorized into two groups: active and passive micromixers [14,15]. Active micromixers require some external source of energy or any moving parts for stimulating the flow and are generally more efficient in mixing than passive micromixers. However, they are difficult to fabricate and integrate with the main microfluidic systems. Passive micromixers avoid such problems by just including geometrical variations in the microchannel to manipulate the laminar flow (uniaxial flow) for mixing of fluids.

There are numerous designs of passive micromixers reported in the literature, working on different mixing principles. The key idea of enhancing the mixing rate is increasing the area of the interface of the fluid streams and decreasing the length of diffusion. Effective micromixers have the capability of increasing the interface area by adopting certain techniques. Chaotic micromixer using staggered herringbone grooves on the wall showed high effectiveness by exponentially increasing the interface area of the fluid streams [16]. Three-dimensional serpentine [17], curved [18–20] and zig-zag [21,22] microchannels have been demonstrated in mixing of fluids. The obstacles in the microchannel have been reported to increase the mixing of fluids by creating advection mixing. Liu et al. fabricated specific bas-relief microstructures on the floor of the microchannel using a direct printing process for mixing [23]. The other technique to increase the interface area of the fluid streams is the multi-lamination of fluid streams by repeatedly splitting and recombining the channels [24,25]. This technique utilizes the two mixing mechanisms: decreasing the length of diffusion and increasing the interface area of fluid streams.

The basic layout of all micromixers can be divided into four separate major units: inlet channels, T- or Y-shaped junctions, mixing channels and outlet channels. Fluid streams are brought into contact by inlet channels connected to a micromixer. Mixing slowly starts right after the fluid streams come into contact at the junction, while major mixing takes place in the mixing channel. Most researchers have focused on designing the mixing channel to create specific flow patterns to enhance mixing. The T- or Y-shaped channel itself is also the simplest T-mixer [26–37]. For example, the simple T-mixer is suited for studying the basic concept on the mechanism of the mixing of fluids.

In T-mixers, researchers have identified three different flow regimes developed at different Reynolds numbers and their influences on mixing [26–35]. These three flow regimes are: stratified, vortex and engulfment flows. In the stratified flow regime, the streamlines of the incoming fluid streams smoothly make a turn and flow along the wall of the microchannel at the bend of the junction forming the symmetric interface in the middle of the main microchannel. In such a situation, mixing is mainly governed by diffusion of molecules across the interface of the fluid streams. The mixing performance is poor because diffusion is a very slow process. At some higher Reynolds numbers (Re $\geq$ 136) [36,37], the vortex flow starts with the formation of the two vortices, and mixing performance is improved due to the increase in the interface area of the fluid streams. The third and the most important regime for attaining effective mixing is the engulfment flow where streamlines intertwine with each other. In this flow regime, the last two physical phenomena contribute to mixing: an increase in the interface area and a decrease in the diffusion length due to the intertwining of the streamlines. Such flow structures are desired to attain rapid mixing. However, such a flow regime starts at a quite high Reynolds number (~200), which would require high pumping power. Furthermore, the fluid streams have less residence time in the channel for proper mixing.

From the above discussions, it becomes obvious that for the range of Reynolds number (Re $\leq$ 136), the fluid streams from the inlet channels make a clear, vertical, thin and straight interface across which mixing takes place. The interface gradually becomes broader as the fluid flows towards the other end of the microchannel. Some efforts have to be done for design modification of a simple T-mixer for an early development of the mixing, inducing complex flows at lower Reynolds numbers. An effort was made for the modification to the simple T-mixer with the objective of enhancing the mixing of fluids named as the vortex micro T-mixer [38].

The vortex micro T-mixer shows the formation of vortex flows at low Reynolds numbers [38]. The generated vortex flow stretches the interface of the fluid streams, which increases the mixing of fluids. In this paper, we present a comparative analysis of simple and vortex micro T-mixers at different Reynolds numbers (1–80), both experimentally and numerically. The study has been further extended to demonstrate the mixing performances of novel designs: planar serpentine microchannels combined with the simple and the vortex T-junctions. The design of the vortex T-mixer was proven to be very promising and hence can be selected for integration with all possible current and future micromixer designs.

2. Physical and Numerical Model

The basic concept of increasing the mixing of fluids has utilized the vortex flow formed by the fluid streams flowing through non-aligned inlet channels into the mixing channel (See Figure 1). Schematics of the physical models of simple and vortex micro T-mixers are shown in Figure 2. The width (200 μm), height (200 μm) and length (5 mm) of the main microchannel are equal and fixed for both the vortex and simple T-mixers. The main focus of the present micromixer study is at the junction of the inlet channels and the mixing channel. For the simple T-mixer, the height of the inlet channels is the same as the height of the mixing channel, while in case of the vortex T-mixer, the height of the inlet channels is half the main microchannel ($h = H/2$). The cross-sectional area of the inlet channels was kept constant and equal for both the vortex (100 μm × 100 μm) and the simple T-mixers (50 μm × 200 μm) in order to compare the mixing performances at similar operating conditions. The length of the inlet channels for both the simple and vortex T-mixers was 1000 μm.

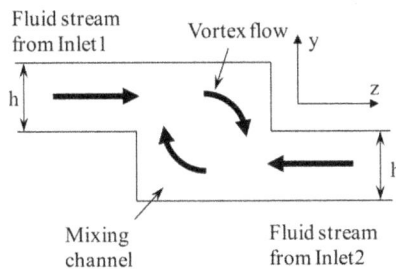

Figure 1. Basic schematic of the idea of vortex micro T-mixer. The two inlet channels of the vortex micro T-mixer are joining the mixing channel at an offset position. The height of the inlet channels is half the height of the mixing channel ($h = H/2$). The two fluid streams from Inlet 1 and Inlet 2 enter the channel at an offset position and create a vortex flow.

The numerical study on the simple and vortex T-mixers has been carried out at different Reynolds numbers, in a range of 1–80. The effects of the two types of T-junction on the mixing performance have been also investigated by integrating them with a planar serpentine microchannel. The Reynolds number has been calculated considering the characteristic dimensions of the main microchannel with the properties of water.

Numerical simulation on both the simple and vortex T-mixers has been carried out by solving continuity, Navier–Stokes and convection-diffusion equations for the mixing of fluids using a finite volume solver, ANSYS CFX 15.0 (ANSYS, Inc., Canonsburg, PA, USA) [39]. The flows calculated in this work were assumed steady and incompressible. The boundary conditions were normal velocity components at the inlets and zero average static pressure at the outlet. The walls were assigned the no-slip boundary condition. Water and ethanol were selected as the two working fluids with all properties taken at 20 °C. The solution was considered to have attained convergence for a root-mean-square value of 10^{-6} for both the mass fraction and momentum.

Structured grids were applied to discretize the computational domain. In order to capture the complex flow field of vortex flow and the deformation of the interfaces of the fluid streams, the sizes of the computational cells were kept small (~4 μm) near the junction. However, downstream of the T-junction, the flow structure becomes simple in the straight microchannel. Therefore, in this zone, the mesh density was kept low, and hence, the cells were not cubical, rather aligned in the direction of the flow.

The mixing performance of the device was evaluated by calculating the mixing index. It is based on the variance of the mass fraction from the mean concentration. The variance was calculated on the plane perpendicular to the x-axis. The values of the mass fraction at equally-spaced sample points on the plane were evaluated to calculate the variance. The mixing index has been defined as:

$$M = 1 - \sqrt{\frac{\sigma^2}{\sigma_{max}^2}} \tag{1}$$

whereas σ^2 is the variance of the mass fraction of the fluid component of the mixture and σ_{max}^2 is the maximum variance.

$$\sigma^2 = \frac{1}{N} \sum (c_i - \bar{c}_m)^2 \tag{2}$$

The values of the mass fraction have been evaluated at N sampling points on the plane. c_i is the mass fraction at sampling point i, and $\bar{c}_m$ is the optimal mixing mass fraction. The value of optimal mass fraction $\bar{c}_m$ is 0.5 for completely mixed fluids.

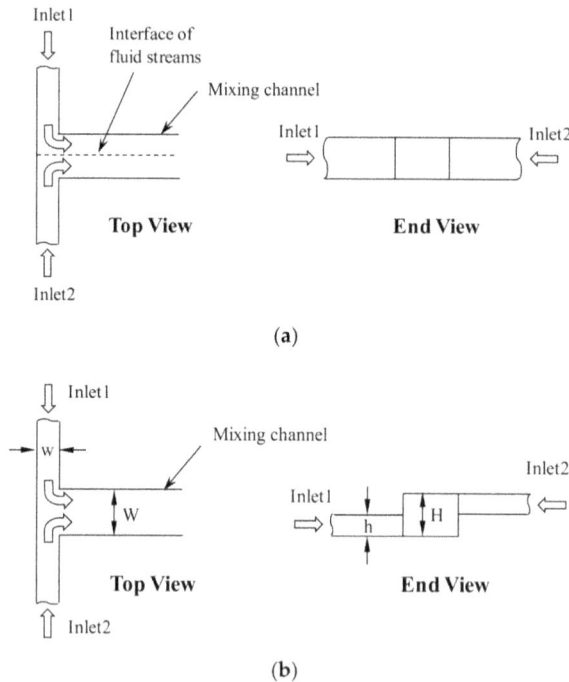

Figure 2. Schematic of the basic idea of the micromixer: (**a**) simple T-mixer and (**b**) vortex micro T-mixer.

3. Device Fabrication and Experiment

3.1. Fabrication

The micromixer was fabricated in PDMS using the soft lithography technique. The polydimethylsiloxane (PDMS) is a silicon-based polymer (colorless, viscous liquid) widely used for fabricating microfluidic devices. Since the fabrication technique using soft lithography is standardized and well known, the various steps are briefly explained here. The replica process used in the present work is described in detail in previous papers [38,40]. SU-8 photoresist (GM1075, Gersteltec Sarl, Pully, Switzerland) was spin coated on a 4-inch silicon wafer using a spin coater to obtain a 100 μm-thick layer. Soft and hard baking was performed on a hot plate. The coated Si-wafer was exposed to UV light using a standard UV aligner through a high resolution dark film mask followed by post baking.

The master mold was developed using SU-8 developer by taking off the unexposed SU-8. The device was fabricated by pouring silicon elastomer and curing agent in a 10:1 weight ratio into the patterned wafer followed by degassing in vacuum desiccators (H42050, Bel-Art product, Wayne, NJ, USA) and heat treatment at 72 °C for 2 h in the convection oven (NDO-400, Sanyo, Osaka, Japan). The device for making the vortex T-mixer was fabricated into two layers, while for the simple T-mixer, into a single layer. The two PDMS layers were bonded together by a manual alignment. Ethanol was applied on the PDMS surface of each layer after corona treatment to preserve the bonding properties of the surface for a sufficiently long time required for manual alignment. The bonded device was kept in the oven at 70 °C for 2 h for heat treatment. The single-layered device was bonded to a glass slide after plasma treatment irreversible sealing. Holes were punched into the PDMS to make the inlet and outlet port using a punch (33-31AA-P, Miltex, Plainsboro, NJ, USA). Figure 3 shows the micrograph of the simple T-mixer, the vortex T-mixer and the planar serpentine channel integrated with the vortex T-junction.

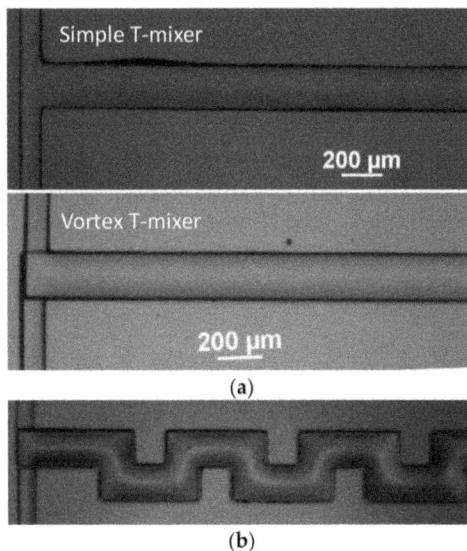

Figure 3. Micrographs of the experimental devices: (**a**) simple and vortex micro T-mixers and (**b**) planar serpentine channel integrated with vortex T-junction.

3.2. Mixing Experiment

A florescent solution was prepared using distilled water (Milli Q purified) and Rhodamine B (Sigma-Aldrich, St. Louis, MO, USA) with a 100 μM concentration. The Rhodamine B was perfectly dissolved in the water using a vortex stirrer (KMC-1300V, Vision Scientific Co., Ltd., Kyeonggi-do, Korea) and ultrasonic waves. It was ensured that the solution was free of any Rhodamine particle, as this may cause coherence in the microchannel, which may locally disorder the distribution of the florescent intensity and hence may influence the results of mixing. The diffusion coefficients of Rhodamine B in water and ethanol are 2.8×10^{-10} m$^2 \cdot$s^{-1} and 1.2×10^{-10} m$^2 \cdot$s^{-1}, respectively, while its molecular weight is 479.17 g/mol [41–44]. Mixing experiments were carried out using a florescent water solution and pure water on an inverted microscope (Ti-u, Nikon, Tokyo, Japan). Planar mixing images were captured using a charge-coupled device (CCD) camera (DS-Qi1Mc, Nikon, Tokyo, Japan). The cross-sectional images were captured using a confocal microscope (LSM10META, Carl Zeiss, Oberkochen, Germany). Fluids were fed into the micromixer by a multi-feed pump (Model 200, KDScience Inc., Holliston, MA, USA) using two syringes through inlet holes connected with Teflon tubes. A constant flow rate of both fluids was maintained in the micromixer for the corresponding Reynolds numbers.

4. Results and Discussion

4.1. Simple and Vortex Micro T-Mixers

An experimental and a numerical study have been carried out on the simple and vortex micro T-mixer for a wide range of Reynolds numbers (1–80). Their mixing performances have been compared for the same operating conditions. In the numerical study, a preliminary test for grid independency has been performed in order to ensure a mesh-independent solution. Five grid systems have been selected with the number of nodes ranging from 3×10^5–2.2×10^6. Finally, a grid system with the number of nodes 1.26×10^6 was selected for efficiently carrying out further simulations. The mixing index was evaluated at a fixed location ($x = 4.5$ mm) in the microchannel at Re = 80 for each grid. There was a small variation (0.91%) in the mixing index for the two grids, i.e., 5.8×10^5 and 9.1×10^5. The relative change in the mixing index between the grid systems having 5.8×10^5 and 9.1×10^5 nodes was 0.73%. The difference was found to be of a similar order (0.78%) as the difference between the grid systems with node numbers 1.26×10^6 and 2.21×10^6. However, considering the change in the number of nodes, the latter relative change in mixing is quite low. Based on this result of the grid test, with the number of nodes 1.26×10^6, the numerical results were expected to have minimum error. The simulation results also have numerical diffusion, which can be minimized by increasing the mesh density [45,46]. Ansari et al. [45] reported such an analysis on the presence of numerical diffusion by showing the mass fraction distribution for different mesh densities.

Figure 4 shows the streamlines for fluid flow from Inlet 1 for the simple and vortex T-mixers. In the simple T-mixer, the projected streamlines are shown for Re = 40, which represents the development of symmetrical flow in the main microchannel. The two incoming fluid streams form similar flow structures till Re = 80. For the range of the Reynolds number reported in the present work, there is no significant effect on the interface in the simple T-mixer, and hence, there is no influence on mixing performance. The end view of the streamlines is shown at different Reynolds numbers (1, 20, 50 and 80) along with the three-dimensional view (Re = 40). It will be helpful to understand the development of the flow field by visualization and to explain the reason for increasing mixing performance. In the vortex T-mixer, at Re = 1, there is no vortex flow due to very low inertia in the incoming fluid streams. The trajectories of the fluid streams from the two inlet channels are parallel to the walls of the main microchannel. At Re = 20, vortex flow can be visualized, but is not strong enough to make full rotation in the mixing channel. The three-dimensional view of the projected streamlines (for vortex T-mixer at Re = 40) shows that the intensity of the vortex flow is highest near the junction and decreases gradually along the microchannel length. The region affected by the vortex flow increases with the increase in the Reynolds number due to the high inertia force. After a certain distance in the channel, the influence of the vortex flow diminishes.

The analysis results for the simple and vortex T-mixer are represented by the mass fraction distributions on the y-z plane (at a fixed axial distance) at various Reynolds numbers for visualizing mixing. The results of numerical and experimental analyses are compared in Figure 5 for Re = 20 and 40. The flow structures are quite similar; even the kinks formed in the numerical simulation at Re = 20 were clearly seen in the cross-sectional image.

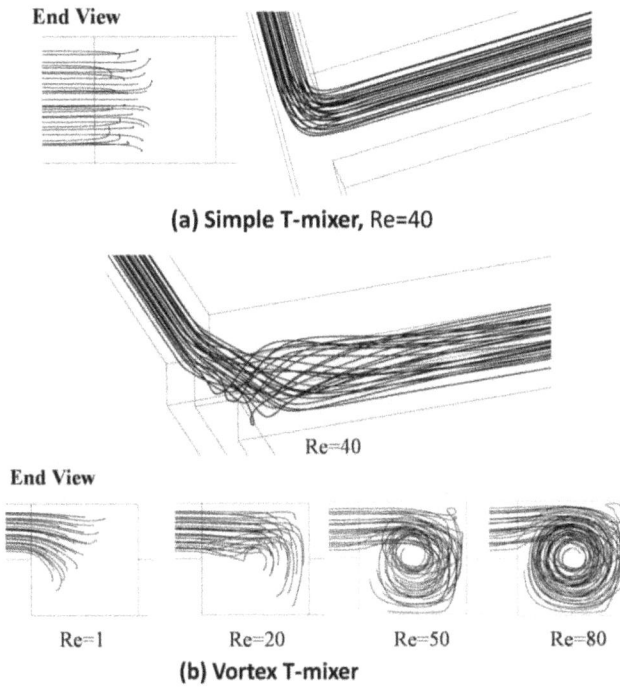

End View

(a) Simple T-mixer, Re=40

Re=40

End View

Re=1   Re=20   Re=50   Re=80

(b) Vortex T-mixer

Figure 4. Projected streamlines shown in end view and three-dimensional view: (**a**) simple micro T-mixer and (**b**) vortex micro T-mixer.

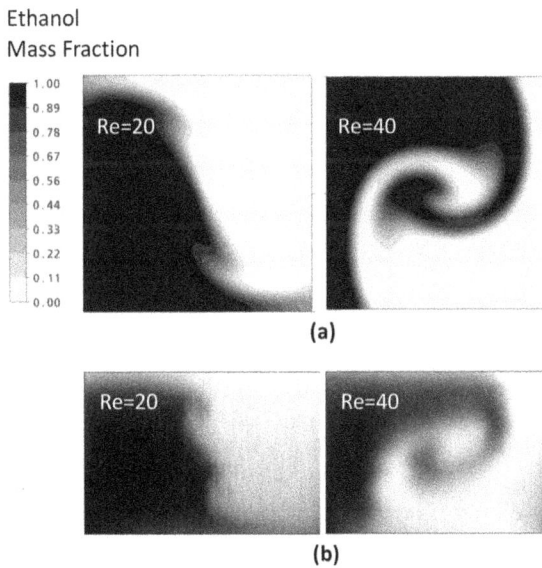

Ethanol
Mass Fraction

Re=20   Re=40

(a)

Re=20   Re=40

(b)

Figure 5. Comparison of the simulation and experimental results for the vortex micro T-mixer: (**a**) numerical results for the mass fraction distribution on y-z plane and (**b**) experimental results for the cross-sectional image [38].

Figure 6a shows the mass fraction distributions for the simple T-mixer on the y-z plane at Re = 1, 40 and 80. The two fluid streams from the inlet channels form a vertical interface in the middle of the microchannel. There is no significant increase in the interface area of the fluid streams till Re = 80. In the vortex T-mixer (Figure 6b), the mass fraction distributions are shown at Re = 1, 30, 50 and 80. In the vortex T-mixer, at Re = 1, the interface of the fluid streams is diagonally oriented to the cross-section of the mixing channel, and there is no formation of any vortex flow due to low inertia force in the fluid streams. The orientation of the interface is similar at all axial locations in the main mixing channel. At Re = 30, the interface of the fluid streams shows some deformation and reorientation. With increasing Reynolds number, the fluid streams from the two inlets strike with high velocity and create stronger vortex flow (Re = 50 and 80). This type of vortex flow, which effectively stretches the interface area of the fluid streams, is desired for effectively increasing mixing. There is a big difference in the flow structures developed in the simple and vortex T-mixers. Hence, there is no improvement in the mixing performance with the increase in Reynolds number. However, in the vortex T-mixer, there is a significant increase in the interface area due to stretching by vortex flow. Such a type of flow field is desired for effectively increasing the interface of the fluid streams and hence attaining better mixing.

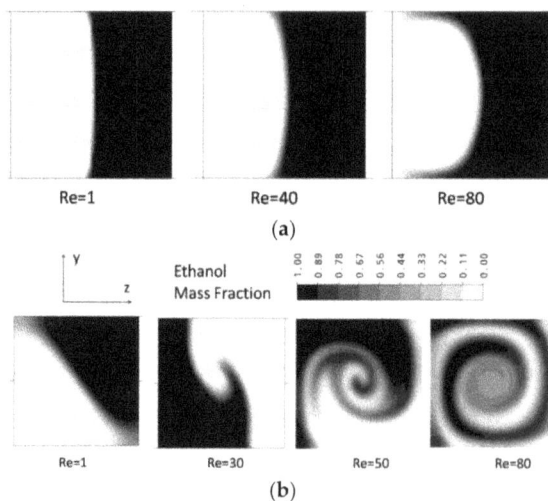

Figure 6. Mass fraction distributions of ethanol on the y-z plane at different Reynolds numbers: (a) simple micro T-mixer and (b) vortex micro T-mixer.

The flow structure and mixing of fluids explained above using the numerical results can be supported by experimental results. Figure 7 shows the planar mixing images obtained by experiment for a simple and a vortex T-mixer at different Reynolds numbers. In the simple T-mixer, the two incoming fluid streams form a clear interface in the middle of the microchannel that exist till the end of the microchannel. The mixing image is presented only for Re = 80, as the flow fields are similar below this Re. On the contrary, the vortex T-mixer shows a varied flow field with the Reynolds number. At Re = 10, a little mixing is observed as the two fluid streams form an interface that exists till the end of the microchannel. The incoming fluid streams are able to make a rotation at Re = 30 and above. However, as the Reynolds number increases, the device shows better mixing performance. At Re = 60 and 80, the distribution of the florescence in the image shows higher uniformity near the exit, showing better mixing.

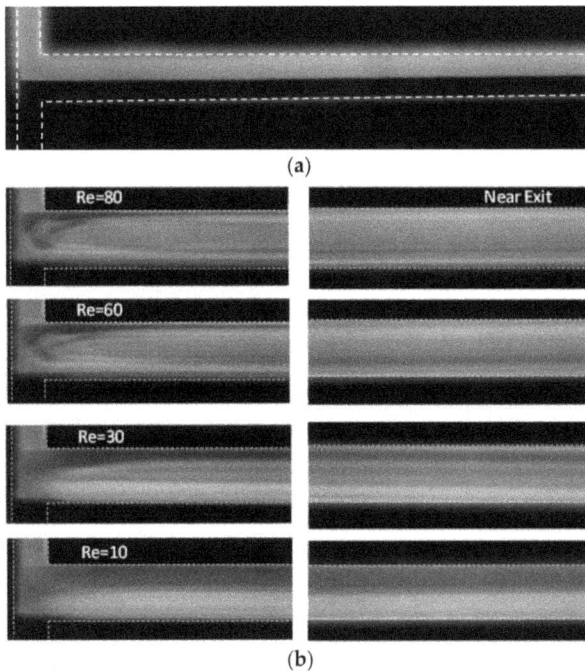

Figure 7. Planar mixing images obtained by experiment for (**a**) the simple T-mixer, Re = 80 and (**b**) the vortex micro T-mixer at different Reynolds numbers near the junction (left) and the exit of the mixing channel (right).

The mixing index has been evaluated along the *x*-axis at different Reynolds numbers (1, 20, 50 and 80) to quantitatively investigate and compare the mixing performances of the simple and vortex T-mixers (see Figure 8). At Re = 1, both the micromixers show similar mixing characteristics, where the mixing performance gradually increases along the channel length. In the initial section of the main mixing channel, the vortex T-mixer shows slightly higher mixing than the simple T-mixer. This may be due to the large interface area of the fluids streams attained by its diagonal position in the mixing channel (see Figure 5b). However, this difference in mixing vanishes in the later part of the channel. The reason for this is not clear, but may be due to some other effects, like diffusion distance, local residence time of the interface of the fluid streams or other fluid dynamics effects. However, at higher Reynolds numbers, inlet channel modification distinguishably affects the mixing performance. In the vortex T-mixers, the rapid jump in mixing is happening in the initial section of the microchannel, which is due to the formation of the vortex flow. Moreover, the flow has no vortex or the flow becomes simple flow in the remaining part of the mixing channel. In the case of the simple T-mixer, very slow mixing is shown, and there is only a small increase in the mixing index from the T-junction to the end of the main microchannel. The simple T-mixer cannot significantly increase the interface area of the fluid for the reported range of Reynolds number (1 ≤ Re ≤ 80). The difference in the mixing performance between the simple and vortex T-mixers increases as the Reynolds number increases due to an increase in the intensity of the vortex flow caused by the high inertia force of the incoming fluid streams.

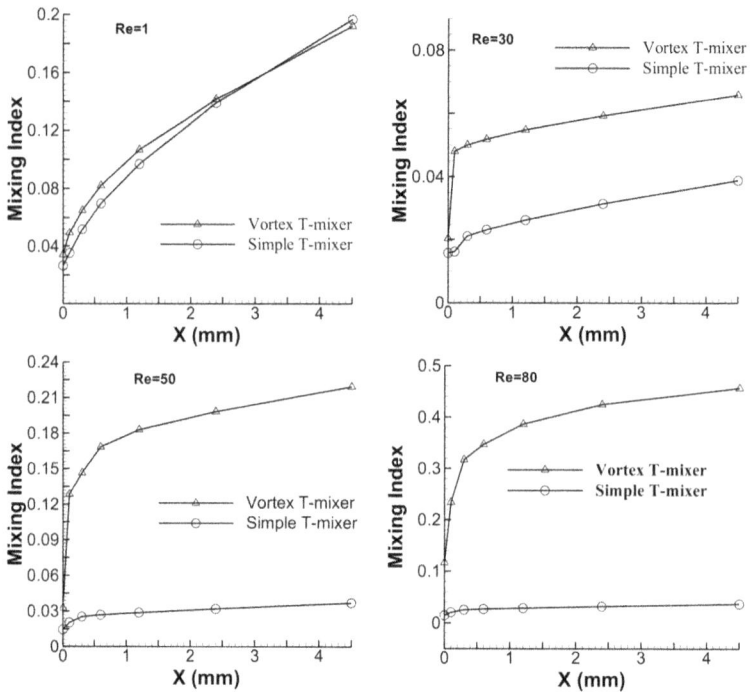

Figure 8. Numerical results for the development of the mixing index along the *x*-axis at various Reynolds numbers for simple and vortex T-mixers with the straight mixing channel.

4.2. Integrating with Serpentine Microchannel

The mixing performances of the simple and vortex T-junctions have been evaluated and compared by combining them with a planar serpentine channel. The contour plots for the mass fraction distribution on the y-z planes are shown for the serpentine microchannel integrated with the simple and vortex T-junctions at Re = 20 and 40 in Figure 9. In the microchannel with the simple T-junction, the interface of the fluid streams seems very clear and shows only little deformation along the *x*-axis. Better mixing performance can be clearly observed for the microchannel integrated with the vortex T-junction. Experimental results showing the planar mixing images for the simple and vortex T-junctions integrated with the serpentine channel are presented near the junction and exit of the channel at Re = 20 and 40 (see Figure 10). The serpentine channel with the simple T-junction shows very little mixing, and the interface of the fluid streams exists till the end of the channel at Re = 20 and 40. The serpentine microchannel with the vortex T-junction shows poor mixing at Re = 20, but shows better mixing at Re = 40. The mixing was quantified by evaluating the mixing index for the planar serpentine channels integrated with the simple and vortex T-junctions at Re = 20 and 40 (see Figure 11). At these Reynolds numbers, higher mixing performances are shown by the serpentine channel with the vortex T-junction as compared to that with the simple T-junction. The difference in the mixing performance between the vortex and simple T-junctions becomes more pronounced at higher Reynolds numbers.

(a)

(b)

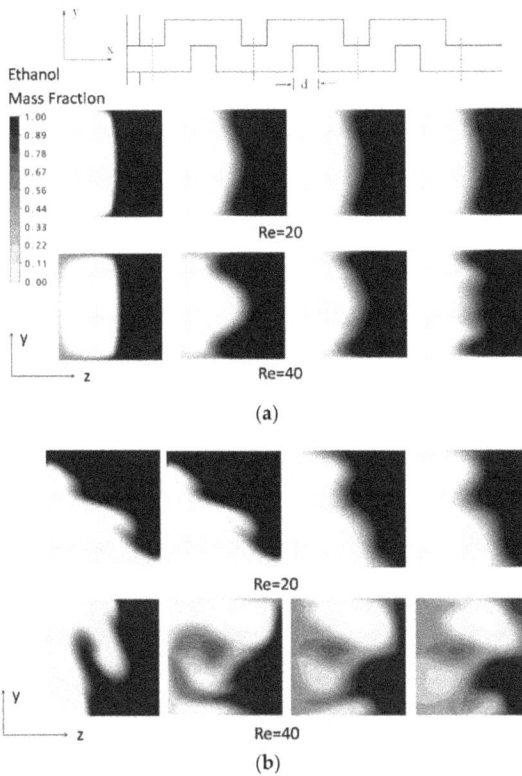

Figure 9. Mass fraction distributions of ethanol on the y-z planes at the initial four axial positions (shown by the dotted lines) in the planar serpentine channel at Re = 20 and 40 integrated with the (**a**) simple micro T-mixer and (**b**) vortex micro T-mixer.

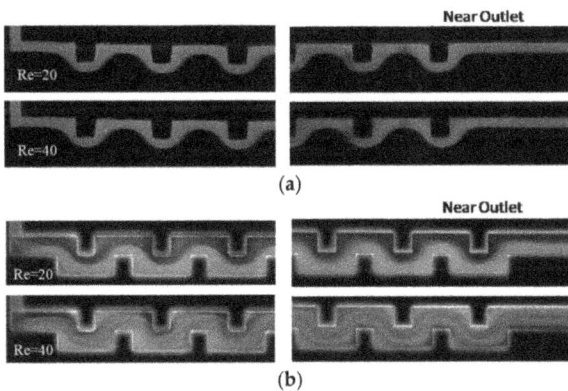

(a)

(b)

Figure 10. Experimental results. Micrographs for the intensity distribution in serpentine channels (at Re = 20 and 40) integrated with the (**a**) simple T-junction and (**b**) vortex T-junction. In the case of the serpentine channel with the simple T-mixer, the height of the channel is equal to the height of the inlet channels. In such a situation, the heights of the serpentine channel with the simple and vortex T-junctions are 100 μm and 200 μm, respectively.

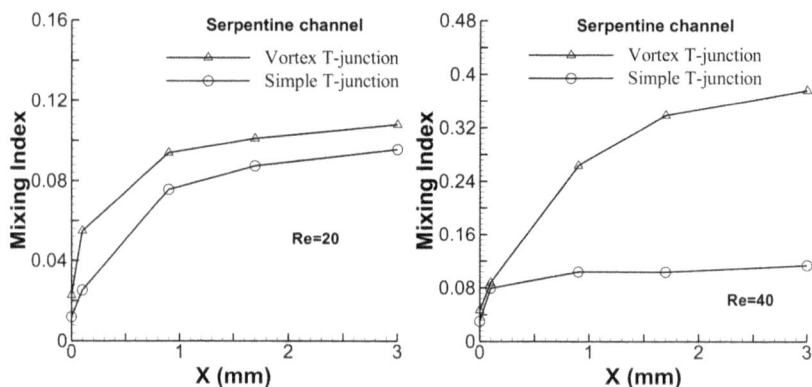

Figure 11. Comparison of the mixing performances of planar serpentine channels integrated with simple and vortex T-junctions.

5. Conclusions

A comparative analysis has been carried out on simple and vortex micro T-mixers, both numerically and experimentally. In the vortex T-mixer, the height of the inlet channels is half the height of the main microchannel. The vortex T-mixer creates vortex flow, which increases the interface area of the fluid streams and, hence, increases mixing performance. The mixing performance of the vortex T-mixers is significantly higher than the simple T-mixers. The flow field developed in the simple T-mixer for the reported range of Reynolds numbers (1–80) is not favorable in facilitating rapid mixing. However, in the vortex T-mixer, mixing performance significantly increases in this range of Reynolds numbers. Mixing performances of the simple and vortex T-junctions have been further demonstrated by integrating them with a planar serpentine microchannel at Re = 20 and 40. The planar serpentine microchannel with the vortex T-junction shows higher mixing performance compared to that with the simple T-junction. Therefore, the vortex T-junction can be selected as an inlet junction of all the possible designs of micromixers to enhance the mixing of fluids.

Author Contributions: Mubashshir Ahmad Ansari performed numerical simulations, carried out experiments, analyzed results and prepared the draft of the paper. Kwang-Yong Kim impeccably revised and improved the manuscript. Sun Min Kim provided technical support to perform experiments and revised the manuscript.

Acknowledgments: This work was supported by the National Research Foundation of Korea (NRF) grants funded by the Korean government (NRF-2016R1A2B4006987). This work was also partially supported by the Inha University WCSL Research Grant, Korea.

Conflicts of Interest: The authors declare no conflict of interest.

References

1. Schwalbe, T.; Autze, V.; Wille, G. Chemical Synthesis in Microreactors. *Chimia* **2003**, *34*, 636–646. [CrossRef]
2. Chow, A.W. Lab-on-a-chip: Opportunities for chemical engineering. *AIChE J.* **2002**, *48*, 1590–1595. [CrossRef]
3. Kopp, M.U.; Mello, A.J.D.; Manz, A. Chemical Amplification: Continuous-Flow PCR on a Chip. *Science* **1998**, *280*, 1046–1048. [CrossRef] [PubMed]
4. Cama, J.; Chimerel, C.; Pagliara, S.; Javera, A.; Keyser, U.F. A label-free microfluidic assay to quantitatively study antibiotic diffusion through lipid membranes. *Lab Chip* **2014**, *14*, 2303–2308. [CrossRef] [PubMed]
5. Lin, Y.C.; Yang, C.C.; Huang, M.Y. Simulation and experimental validation of micro polymerase chain reaction chips. *Sens. Actuators B Chem.* **2000**, *71*, 127–133. [CrossRef]
6. Weigl, B.H.; Bardell, R.L.; Cabrera, C.R. Lab-on-a-chip for drug development. *Adv. Drug Deliv. Rev.* **2003**, *55*, 349–377. [CrossRef]

7. Dittrich, P.S.; Manz, A. Lab-on-a-chip: Microfluidics in drug discovery. *Nat. Rev. Drug Discov.* **2006**, *5*, 210–218. [CrossRef] [PubMed]
8. Freitas, S.; Walz, A.; Merkle, H.P.; Gander, B. Solvent extraction employing a static micromixer: A simple, robust and versatile technology for the microencapsulation of proteins. *J. Microencapsul.* **2003**, *20*, 67–85. [CrossRef] [PubMed]
9. Stone, H.A.; Stroock, A.D.; Ajdari, A. Engineering flows in small devices: Microfluidics toward a lab-on-a-chip. *Annu. Rev. Fluid Mech.* **2004**, *36*, 381–411. [CrossRef]
10. Gossett, D.R.; Weaver, W.M.; Mach, A.J.; Hur, S.C.; Tse, H.T.K.; Lee, W.; Amini, H.; Carlo, D.D. Label-free cell separation and sorting in microfluidic systems. *Anal. Bioanal. Chem.* **2010**, *397*, 3249–3267. [CrossRef] [PubMed]
11. Gossett, D.R.; Tse, H.T.K.; Lee, S.A.; Ying, Y.; Lindgren, A.G.; Yang, O.O.; Rao, J.; Clark, A.T.; Carlo, D.D. Hydrodynamic stretching of single cells for large population mechanical phenotyping. *Proc. Natl. Acad. Sci. USA* **2012**, *109*, 7630–7635. [CrossRef] [PubMed]
12. Zilionis, R.; Nainys, J.; Veres, A.; Savova, V.; Zemmour, D.; Klein, A.M.; Mazutis, L. Single-cell barcoding and sequencing using droplet microfluidics. *Nat. Protoc.* **2017**, *12*, 44–73. [CrossRef] [PubMed]
13. Pagliara, S.; Dettmer, S.L.; Keyser, U.F. Channel-Facilitated Diffusion Boosted by Particle Binding at the Channel Entrance. *Phys. Rev. Lett.* **2014**, *113*, 048102. [CrossRef] [PubMed]
14. Nguyen, N.T.; Wu, Z.G. Micromixers—A review. *J. Micromech. Microeng.* **2005**, *15*, R1–R16. [CrossRef]
15. Hessel, V.; Lowe, H.; Schonfeld, F. Micromixers—A review on passive and active mixing principles. *Chem. Eng. Sci.* **2005**, *60*, 2479–2501. [CrossRef]
16. Stroock, A.D.; Dertinger, S.K.W.; Ajdari, A.; Mezic, I.; Stone, H.A.; Whitesides, G.M. Chaotic mixer for microchannels. *Science* **2002**, *295*, 647–651. [CrossRef] [PubMed]
17. Liu, R.H.; Stremler, M.A.; Sharp, K.V.; Olsen, M.G.; Santiago, J.G.; Adrian, R.J.; Aref, H.; Beebe, D.J. Passive mixing in a three-dimensional serpentine microchannel. *J. Microelectromech. Syst.* **2000**, *9*, 190–197. [CrossRef]
18. Jiang, F.; Drese, K.S.; Hardt, S.; Kupper, M.; Schonfeld, F. Helical flows and chaotic mixing in curved micro channels. *AIChE J.* **2004**, *50*, 2297–2305. [CrossRef]
19. Schonfeld, F.; Hardt, S. Simulation of helical flows in microchannels. *AIChE J.* **2004**, *50*, 771–778. [CrossRef]
20. Yamaguchi, Y.; Takagi, F.; Yamashita, K.; Nakamura, H.; Maeda, H.; Sotowa, K.; Kusakabe, K.; Yamasaki, Y.; Morooka, S. 3-D simulation and visualization of laminar flow in a microchannel with hair-pin curves. *AIChE J.* **2004**, *50*, 1530–1535. [CrossRef]
21. Chen, J.K.; Yang, R.J. Electroosmotic flow mixing in zigzag microchannels. *Electrophoresis* **2007**, *28*, 975–983. [CrossRef] [PubMed]
22. Mengeaud, V.; Josserand, J.; Girault, H.H. Mixing processes in a zigzag microchannel: Finite element simulations and optical study. *Anal. Chem.* **2002**, *74*, 4279–4286. [CrossRef] [PubMed]
23. Liu, A.L.; He, F.Y.; Wang, K.; Zhou, T.; Lu, Y.; Xia, X.H. Rapid method for design and fabrication of passive micromixers in microfluidic devices using a direct-printing process. *Lab Chip* **2005**, *5*, 974–978. [CrossRef] [PubMed]
24. Kim, D.S.; Lee, S.H.; Kwon, T.H.; Ahn, C.H. A serpentine laminating micromixer combining splitting/recombination and advection. *Lab Chip* **2005**, *5*, 739–747. [CrossRef] [PubMed]
25. Wu, Z.G.; Nguyen, N.T. Convective-diffusive transport in parallel lamination micromixers. *Microfluid. Nanofluid.* **2005**, *1*, 208–217. [CrossRef]
26. Bothe, D.; Sternich, C.; Warnecke, H.J. Fluid mixing in a T-shaped micro-mixer. *Chem. Eng. Sci.* **2006**, *61*, 2950–2958. [CrossRef]
27. Wong, S.H.; Ward, M.C.L.; Wharton, C.W. Micro T-mixer as a rapid mixing micromixer. *Sens. Actuators B Chem.* **2004**, *100*, 359–379. [CrossRef]
28. Soleymani, A.; Kolehmainen, E.; Turunen, I. Numerical and experimental investigations of liquid mixing in T-type micromixers. *Chem. Eng. J.* **2008**, *135*, S219–S228. [CrossRef]
29. Soleymani, A.; Yousefi, H.; Turunen, I. Dimensionless number for identification of flow patterns inside a T-micromixer. *Chem. Eng. Sci.* **2008**, *63*, 5291–5297. [CrossRef]
30. Adeosun, J.T.; Lawal, A. Numerical and experimental studies of mixing characteristics in a T-junction microchannel using residence-time distribution. *Chem. Eng. Sci.* **2009**, *64*, 2422–2432. [CrossRef]
31. Mendels, D.A.; Graham, E.M.; Magennis, S.W.; Jones, A.C.; Mendels, F. Quantitative comparison of thermal and solutal transport in a T-mixer by FLIM and CFD. *Microfluid. Nanofluid.* **2008**, *5*, 603–617. [CrossRef]

32. Ma, Y.B.; Sun, C.P.; Fields, M.; Li, Y.; Haake, D.A.; Churchill, B.M.; Ho, C.M. An unsteady microfluidic T-form mixer perturbed by hydrodynamic pressure. *J. Micromech. Microeng.* **2008**, *18*, 045015. [CrossRef] [PubMed]

33. Engler, M.; Kockmann, N.; Kiefer, T.; Woias, P. Numerical and experimental investigations on liquid mixing in static micromixers. *Chem. Eng. J.* **2004**, *101*, 315–322. [CrossRef]

34. Kockmann, N.; Kiefer, T.; Engler, M.; Woias, P. Convective mixing and chemical reactions in microchannels with high flow rates. *Sens. Actuators B Chem.* **2006**, *117*, 495–508. [CrossRef]

35. Dreher, S.; Kockmann, N.; Woias, P. Characterization of laminar transient flow regimes and mixing in T-shaped micromixers. *Heat Transf. Eng.* **2009**, *30*, 91–100. [CrossRef]

36. Kastner, J.; Kockmann, N.; Woias, P. Convective mixing and reactive precipitation of barium sulfate in microchannels. *Heat Transf. Eng.* **2009**, *30*, 148–157. [CrossRef]

37. Thomas, S.; Ameel, T.; Guilkey, J. Mixing kinematics of moderate Reynolds number flows in a T-channel. *Phys. Fluids* **2010**, *22*, 031601. [CrossRef]

38. Ansari, M.A.; Kim, K.Y.; Anwar, K.; Kim, S.M. Vortex micro T-mixer with non-aligned inputs. *Chem. Eng. J.* **2012**, *181*, 846–850. [CrossRef]

39. ANSYS. *Solver Theory Guide*; CFX-15.0; ANSYS Inc.: Canonsburg, PA, USA, 2013.

40. Ansari, M.A.; Kim, K.Y.; Anwar, K.; Kim, S.M. A novel passive micromixer based on unbalanced splits and collisions of fluid streams. *J. Micromech. Microeng.* **2010**, *20*, 055007. [CrossRef]

41. Johnson, T.J.; Ross, D.; Locascio, L.E. Rapid microfluidic mixing. *Anal. Chem.* **2002**, *74*, 45–51. [CrossRef] [PubMed]

42. Culbertson, C.T.; Jacobson, S.C.; Ramsey, J.M. Dispersion sources for compact geometries on microchips. *Anal. Chem.* **1998**, *70*, 3781–3789. [CrossRef]

43. Gunthur, A.; Khan, S.A.; Thalmann, M.; Thrachsel, F.; Jensen, K.F. Transport and reaction in segmented gas-liquid flow. *Lab Chip* **2004**, *4*, 278–286. [CrossRef] [PubMed]

44. Kim, D.S.; Lee, S.W.; Kwon, T.H.; Lee, S.S. A barrier embedded chaotic micromixer. *J. Micromech. Microeng.* **2004**, *14*, 798–805. [CrossRef]

45. Hardt, S.; Schönfeld, F. Laminar mixing in different interdigital micromixers: II. Numerical simulations. *AIChE J.* **2003**, *49*, 578–584. [CrossRef]

46. Ansari, M.A.; Kim, K.Y. Parametric study on mixing of two fluids in a three-dimensional serpentine microchannel. *Chem. Eng. J.* **2009**, *146*, 439–448. [CrossRef]

micromachines

MDPI

Article

A Numerical Research of Herringbone Passive Mixer at Low Reynold Number Regime

Dongyang Wang [1], Dechun Ba [1], Kun Liu [1,*], Ming Hao [1], Yang Gao [2], Zhiyong Wu [3] and Qi Mei [4,*]

[1] School of Mechanical Engineering and Automation, Northeastern University, Shenyang 110819, China; wdysend@gmail.com (D.W.); dcba@mail.neu.edu.cn (D.B.); 1500445@stu.neu.edu.cn (M.H.)
[2] Silicon Steel Department, Baosteel Co., Ltd., Shanghai 201900, China; 702172@baosteel.com
[3] Research Center for Analytical Sciences, Northeastern University, Shenyang 110819, China; zywu@mail.neu.edu.cn
[4] Department of Oncology, Tongji Hospital, Tongji Medical College, Huazhong University of Science and Technology, Wuhan 4330030, China
* Correspondence: kliu@mail.neu.edu.cn (K.L.); boris.meiqi@gmail.com (Q.M.); Tel.: +86-24-8367-6945 (K.L.); +86-27-6937-8806 (Q.M.)

Received: 28 September 2017; Accepted: 29 October 2017; Published: 31 October 2017

Abstract: Passive mixing based on microfluidics has won its popularity for its unique advantage, including easier operation, more efficient mixing performance and higher access to high integrity. The time-scale and performance of mixing process are usually characterized by mixing quality, which has been remarkably improved due to the introduction of chaos theory into passive micro mixers. In this paper, we focus on the research of mixing phenomenon at extremely low Reynold number (Re) regime in a chaotic herringbone mixer. Three-dimensional (3D) modeling has been carried out using computational fluid dynamics (CFD) method, to simulate the chaos-enhanced advection diffusion process. Static mixing processes using pressure driven and electric field driven modes are investigated. Based on the simulation results, the effects of flow field and herringbone pattern are theoretically studied and compared. Both in pressure driven flow and electro-osmotic flow (EOF), the mixing performance is improved with a lower flow rate. Moreover, it is noted that with a same total flow rate, mixing performance is better in EOF than pressure driven flow, which is mainly due to the difference in flow field distribution of pressure driven flow and EOF.

Keywords: passive mixing; mixing quality; herringbone pattern; extremely low Re; electro-osmotic flow; pressure driven flow; CFD simulation

---

1. Introduction

Microfluidic technique is a discipline with widespread use and rapid development [1]. With the rapid development of Lab-on-a-Chip (LOC), integrated microfluidic systems have shown great prospect due to its unique advantages, and are widely used in biochemical analysis [2,3], diagnosis detection [4], analytical chemistry [5], logical operation [6] and drug delivery [7], etc.

As an essential component, rapid and efficient mixing acts the role of mixing two or more samples for analysis [8]. Micro mixing based on microfluidics platforms, is widely considered as though an efficient approach [9]. Conventional micro mixers are classified as active mixers and passive mixers [10,11]. In active micromixers, external energy sources, such as electrical [12], magnetic [13], and sound fields [14] are required to develop the mixing process, which makes the operation of mixing skilled-required and difficult to control.

In passive mixers, no external energy sources are necessary. To develop mixing process efficiently, complex channel geometries and heterogeneous channel surface are utilized to increase the contact interface and decrease the diffusion distance, thus, to enhance the mixing performance [11]. With increasing need for better mixing performance, various passive mixers with novel architecture are proposed, such as butterfly mixers [15], Zigzag mixers [16], over bridged mixers [17], and herringbone mixers [18], etc. Since Stroock et al. first proposed a chaotic mixer with herringbone structure [18], passive micro mixer with herringbone pattern has become the most popular design concept due to its simple setup and robust operation. In herringbone mixers, cycled herringbone is patterned on the wall along the main channel. Based on the herringbone pattern, diffusion and advection are enhanced significantly [19–21].

To optimize the mixing performance in passive mixers, numbers of simulation work have been carried out. Computational fluid dynamics (CFD) method has been widely used to investigate mixing mechanism, yielding a better comprehension of the chaotic enhancement effects by the herringbone grooved patterns [22–24]. Using CFD and particle tracking methods, the effects of geometric parameters such as groove depth, number of grooves per cycle and groove width on the mixing quality. The effect of the groove asymmetry and the number of grooves per half cycle on the mixing performance has been also investigated easily, using CFD modelling coupled with Lattice-Boltzmann method [25].

In our previous experimental work, we conduct passive mixing a herringbone mixer at extremely low flow rate. Passive mixing has been developed efficiently in both pressure driven flow and electroosmotic flow. It was observed that with the same total flow rate, passive mixing driven in electroosmotic flow shows a better mixing performance.

In this paper, we aim to study the passive mixing numerically, and novel propose a numerically study to compare the mixing under different driven modes. Specifically, we mainly focus on the effect of flow field on mixing performance in a passive mixer model with herringbone structure under different mixing modes: mixing in pressure driven flow and electroosmotic flow. Using CFD method, we conduct a numerical research on mass transfer processes inside the micro mixer. The effects of flow rate, direct current (DC) driven electric potential and herringbone structure on concentration distribution have been analyzed theoretically. Within same geometry and total flow rate, we further compare the mixing quality to disclose the reason of difference in mixing quality.

2. Problem Formation

The problem originates from our previous experimental work [26]. In a herringbone micro mixer, passive mixing was developed efficiently in pressure driven flow and electroosmotic flow (EOF), respectively. It was observed that with a same total flow rate, mixing performance is better in electroosmotic flow than that in pressure driven flow.

In this work, we aim to elaborate mixing performance difference in these two driven modes theoretically, using numerical simulation. The mixing processes are simulated numerically in both pressure driven flow and electroosmotic flown with a three-dimensional (3D) herringbone structure at low *Re* regime. Based on the experimental results, we first simplify the mixer geometry to 21 mm in length, at which length a fully well-developed mixing performance was obtained experimentally. The design of the model is detailed demonstrated in Appendix A.

Figure 1 is schematic of the physical models in this work. In a straight main channel, herringbone pattern is designed on the wall. Pressure driven flow is driven by a syringe pump at the outlet, while a DC electric field is applied to driven the electroosmotic flow. Micro mixing develops along the main channel passively.

Hydrodynamic flow driven by syringe pump

Electro-osmotic flow driven by DC electric field

DC electric
potential

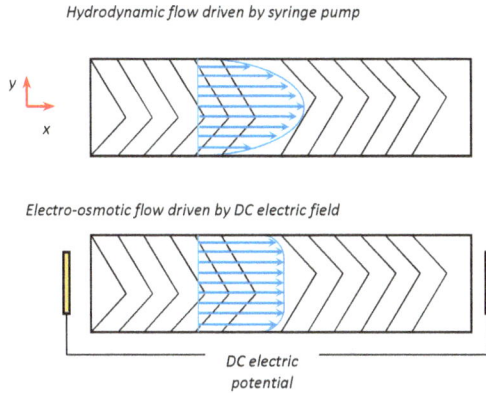

Figure 1. Schematic of two driven modes in this work, hydrodynamic flow and electro-osmotic flow.

3. Theory and Method

3.1. Pressure Driven Flow

When driven hydrodynamically by pressure, the fluid flow in the mixer is governed by continuity equation and Navier-Stokes (N-S) equation. Typical continuity equation is given by:

$$\partial\rho/\partial t + \nabla\cdot(\rho\vec{u}) = 0 \tag{1}$$

While N-S equation is:

$$\partial\vec{u}/\partial t + (\vec{u}\cdot\nabla)\vec{u} = -(1/\rho)\nabla P + (\mu/\rho)\nabla^2\vec{u} + \vec{F} \tag{2}$$

where ρ, $\vec{u}$, P and μ denote the density, velocity vector, pressure and dynamic viscosity of the liquid. Here, $\vec{F}$ represents the force exerted on the flow element at a micro scale, also known as body force. To improve simulation efficiency and obtain fine agreement between numerical calculations and experimental data, the liquid is assumed to be incompressible Newtonian fluids in CFD studies. Hence, the density of the fluid keeps constant, $\partial\rho/\partial t = 0$. Then Equations (1) and (2) can be simplified as follows:

$$\text{Continuity equation}: \nabla(\vec{u}) = 0. \tag{3}$$

Momentum equation (modified Navier-Stokes equation):

$$\rho(D\vec{u}/Dt) = \rho(\partial\vec{u}/\partial t + \vec{u}\cdot\nabla\vec{u}) = -\nabla P + \mu\nabla^2\vec{u} + \vec{F} \tag{4}$$

3.2. Electro-Osmotic Flow (EOF)

The electro-osmotic flow, driven by DC electric field, is assumed to be a Newtonian fluid, and then the motion of an aqueous electrolyte solution in microchannel is governed by the N-S equations. The body force $\vec{F}$ in N-S equation is given by:

$$\vec{F} = \rho_e\cdot\vec{E} \tag{5}$$

Here, $\vec{E}$ is electric field strength and ρ_e is the density of net charge. Then N-S equation is modified into

$$\rho(D\vec{u}/Dt) = \rho(\partial\vec{u}/\partial t + \vec{u}\cdot\nabla\vec{u}) = -\nabla P + \mu\nabla^2\vec{u} + \rho_e\cdot\vec{E} \tag{6}$$

Specifically, the electric field strength is given by the gradient of electrostatic potential Ψ,

$$\vec{E} = \nabla\Psi \tag{7}$$

According to electric double layer (EDL) theory, the electrostatic potential Ψ and the distribution of ions in the solution by Poison Equation:

$$\nabla^2\Psi = -\rho_e/\varepsilon\varepsilon_0 \tag{8}$$

where ε is the dielectric constant of the solution, ε_0 is the permittivity of vacuum. The net charge distribution is stated by Boltzmann theory:

$$\rho_e = -2z_0 en_\infty \sinh(z_0 e\Psi/k_B T) \tag{9}$$

Then the electrostatic potential is expressed by:

$$\nabla^2\Psi = -(2z_0 en_\infty/\varepsilon\varepsilon_0) \sinh(z_0 e\Psi/k_B T) \tag{10}$$

3.3. Mass Transfer

The species concentration in the process of diffusive mass transport is governed by the Fick's first law (one-dimensional form):

$$f = -d(\partial c/\partial x_i) \tag{11}$$

Here, c, f and d are the concentration and species flux and diffusion coefficient respectively. And x_i denotes the diffusion direction, whereas the diffusion always happens in the opposite direction of the concentration gradient. Fick second law describes the temporal behavior of concentration profile, here in one-dimensional form.

Fick second law (one-dimensional form):

$$\partial c/\partial t = d(\partial^2 c/\partial x_i^2) \tag{12}$$

Due to the presence of convection in the diffusive mixing model, an additional convective mass flow with velocity vector $\vec{u}$ extends the Fick second law to the advection-diffusion equation:

$$\partial c/\partial t + (\vec{u}\cdot\nabla)c = d\nabla^2 c. \tag{13}$$

3.4. Statistics and Evaluation

In this study, to quantify the mixing performance, a variance-based method is utilized on cut planes (y-z plane) perpendicular to the x-axis. In view of evaluation with statistics method, the mixing quality α_m of the mixer is described in a discrete form:

$$\alpha_m = 1 - \sqrt{\left[\sum_{i=1}^{N} (c_i - \bar{c})^2\right]/N} \, /\bar{c} \tag{14}$$

where c_i and $\bar{c}$ represent different local concentration and the total mean concentration of the profile in a transfer cross-sectional plane A, which is perpendicular to the downstream direction in the main channel. The term N denotes the population of the sample points in the plane.

3.5. Simulation Method

To properly elaborate chaotic mixing process at a micro scale, Finite Elements Method (FEM) simulations are conducted combined with a commercial software COMSOL Multiphysics 4.4 software (COMSOL Inc., Stockholm, Sweden). This work coupled computational multi-physics, including fluidics, advection diffusion transport and static electric field. Governing equations of the simulations are stated in Appendix B.

A preliminary grid dependence was conducted to minimize the influence of mesh number on the resulting mixing efficiency. Detailed mesh and time step setting are stated in Appendix C. Table 1 shows the constant parameters of the fluid used in this work.

Table 1. Parameters of some variables utilized in this work.

Parameter	Unit	Value
ρ, density	kg/m^3	1×10^3
μ, viscosity	Pa·s	1×10^{-3}
ε_r, relative permittivity	-	8×10
ε_0, permittivity	F/m	8.85×10^{-12}
d, coefficient of diffusion	m^2/s	1×10^{-10}
σ, conductivity	S/m	1.0

The boundary and initial condition is included in Table 2. For simulation of pressure driven flow, a constant flow rate, q_w is exerted at outlet without any extra pressure. In the simulation of electroosmotic flow, a DC electric flied is set between the outlet and the inlet. No-slip wall condition is used in every simulation, while a zeta potential at -0.1 V is applied in the simulation of mixing in EOF. Two inlets introduce the solution and pure water into the main channel respectively. Micro mixing develops once the solution meets water at the start point of main channel. The initial concentration of the solution is 0.1 mol/m^3 in every simulation.

Table 2. Key boundary condition and initial condition applied in the simulations of micro mixing in pressure driven flow and EOF.

Mode	Initial Condition	Boundary Condition	
		Wall condition	Outlet
Pressure driven flow	concentration 0.1 mol/m^3	non-slip	Flow rate, q_w (µL/min)
Electroosmotic flow (EOF)	concentration 0.1 mol/m^3	Wall condition non-slip, insulated, zeta	Outlet Electric potential, E (V)

4. Results and Discussion

4.1. Mixing in Pressure Driven Flow

The first thing to realize is that flow behavior is of vital significance in passive mixing process. In this work, we aim to study the process of mixing with chaotic enhancement at the extreme low Re regime. Reynold number, defined as $Re = \rho u d / \mu$, means the relative ratio of inertial forces to viscous forces. The flow rates in this work and corresponding Re is listed in Table 3. Re is this work peaks at 1.4×10^{-2} under a 5.0 µL/min, indicating that the whole mixing develops at an extremely low Re regime.

Table 3. Various flow rate in this work and relevant Re and outlet velocity.

Flow Rate (µL/min)	0.6	0.9	1.3	2.5	5.0
Velocity (m/s)	8.5×10^{-4}	1.28×10^{-3}	1.85×10^{-3}	3.56×10^{-3}	7.12×10^{-3}
Re	1.6×10^{-3}	2.5×10^{-3}	3.5×10^{-3}	6.8×10^{-3}	1.4×10^{-2}

To profile the flow field in the mixer, the velocity distribution across the main channel is configure and compared in Figure 2. In specific, the middle point of the main channel is taken into consideration.

The red line in Figure 2a represents a series of continuous positions in y direction. While the x and z coordinates are 10.5 mm and 22.5 μm, which represent the middle position in both length and height direction. As can be seen, the velocity fluctuates cross the main channel in width direction, different from typical Poiseuille flow. This is mainly because of introducing of herringbone structure. The enhancing effect of herringbone structure is further investigated numerically.

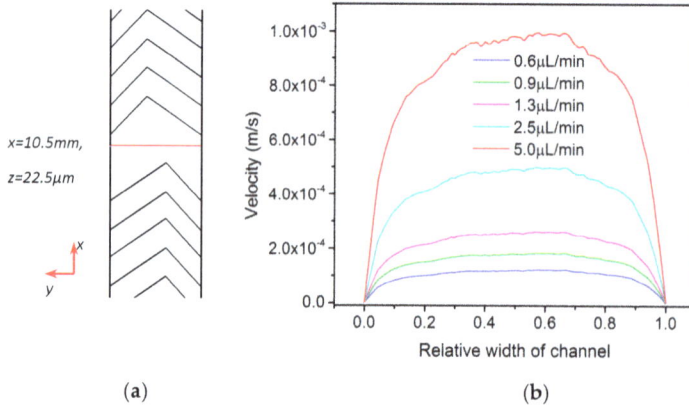

(a)   (b)

Figure 2. (a) Description of the middle position, (b) Velocity comparison at middle position of main channel in pressure driven flow.

Figure 3 compares the enhancing effect of herringbone structure on mixing quality, with a 1.3 μL/min outlet flow rate. As can been seen, the mixing performance is promoted significantly throughout the whole work.

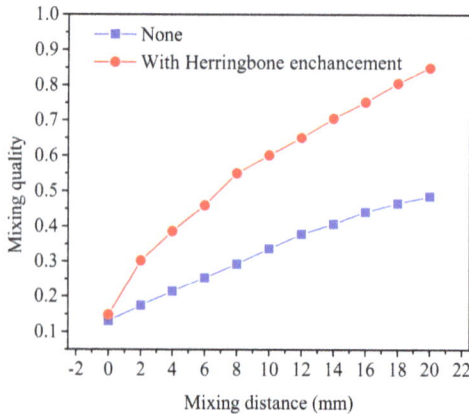

Figure 3. Comparison of mixing quality with and without herringbone structure. Red is the mixing quality along the main channel with herringbone enhancement, while blue is the mixing quality without any enhancement pattern.

The effect of outlet flow rate has been further measured, since it is directly linked to the convective flow throughout the main channel. To visualize the mixing development, the concentration profiles along the channel are demonstrated in Figure 4. At the very beginning of the mixing process ($x = 0$), the concentration distribution is same under different Re. As mixing develops along the channel, species distributes diversely. At outlet($x = 21$ mm), the distribution under a lower *Re* is more uniform than that under a higher *Re*. It is concluded that when the *Re* decreases, the mixing shows a better performance.

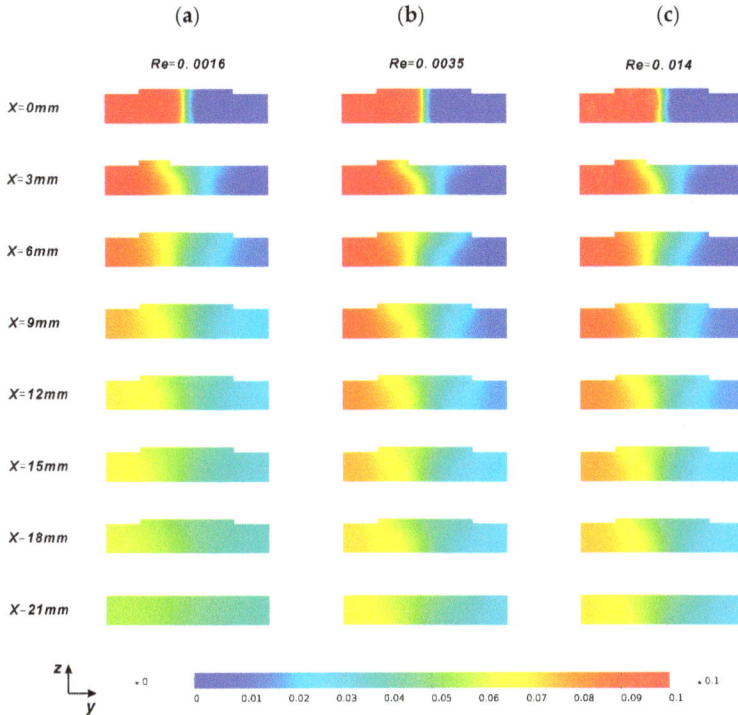

Figure 4. The concentration profiles along the main channel for various *Re*, (**a**) (*Re* = 0.0016); (**b**) (*Re* = 0.0035); (**c**) (*Re* = 0.014). The uniformity of the color represents the mixing performance.

Figure 5 further illustrates the effect of flow rate on mixing performance. Mixing quality along the mixing main channel for different flow rate is compared. In general, the mixing behaves better, if the *Re* goes down. The mixing quality at 20 mm increases from 72.3% to 97.7% as flow rate decreasing from 5.4 µL/min to 0.6 µL/min, which is strictly close to the results in experimental research. According to fluid dynamics, the mixing mechanism comprises convective mixing and diffusive mixing. In the main channel with herringbone patterned on the wall, the initial fluid elements are split and stretched into a multitude of smaller lamellae. Thus, a convective mixing of the species was developed. By the meantime, diffusive mass transfer occurs between the lamellae, which is enhanced by further thinning of lamellae.

The mean residence time was defined as $t_p = L/w$, where L and w denote the length of the channel and the mean velocity in the channel. We note that at very low Reynolds number, the mean residence time is extremely long. A decrease in *Re* leads to a decreased mean velocity and an increasing mean residence time, which in turn promotes the mixing behavior. Thus, we can conclude that in passive mixing at extremely low *Re* regime, molecular diffusion takes place and dominates the mixing processes. Lower flow field ensures a better mixing.

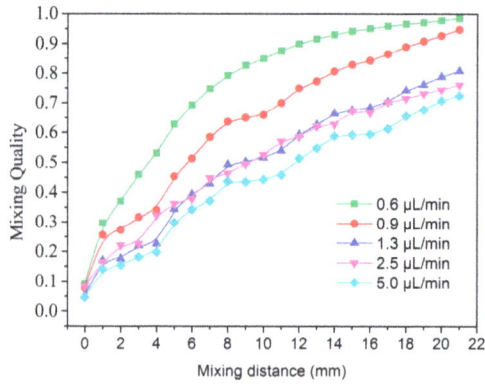

Figure 5. Comparison of the mixing quality along the mixing main channel under various flow rate.

4.2. Mixing Ing Electro-Osmotic Flow (EOF)

In the simulation of the passive mixing in EOF, three different driven electric potential (250 V, 1000 V and 1500 V) are utilized, which is the same as the experimental setup. Similar to study of mixing in pressure driven flow above, we first investigate the flow field behavior. Flow fields are profiled using the velocity in the middle position of the channel in Figure 6, same as the continuous position utilized in the study of mixing in pressure driven flow. With a higher electric potential, the velocity increases correspondingly. However, the velocity fluctuates in y direction, indicating the disturbance of conditions herringbone structure.

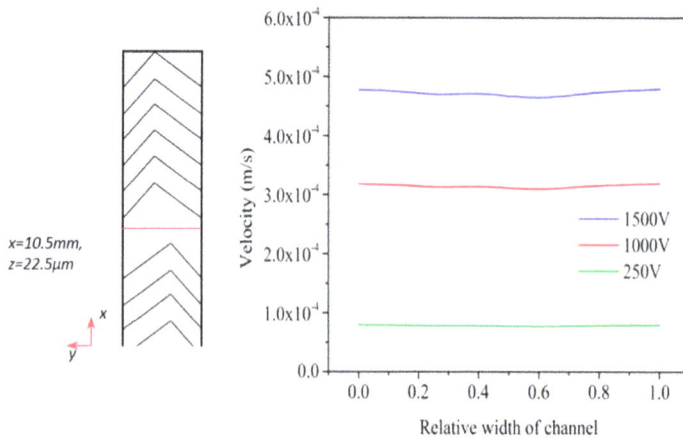

Figure 6. (**a**) Description of the middle position, (**b**) Velocity comparison at middle position of main channel in EOF.

Then the effect of flow rate on mixing performance is qualified in Figure 7. It is noted that the mixing performs better under a lower electric potential, the mixing performance is improved from 0.83 to 0.97. According to the discussion above, at very low Reynolds number, the mean residence time is extremely long. A lower electric potential provides a slower flow field, and reduces *Re*. A decrease in *Re* leads to a decreased mean velocity and an increasing mean residence time, which in turn promotes the mixing behavior. Thus, we can conclude that molecular diffusion takes place and dominates the passive mixing in electroosmotic flow at extremely low *Re* regime. Lower flow field ensures a better mixing.

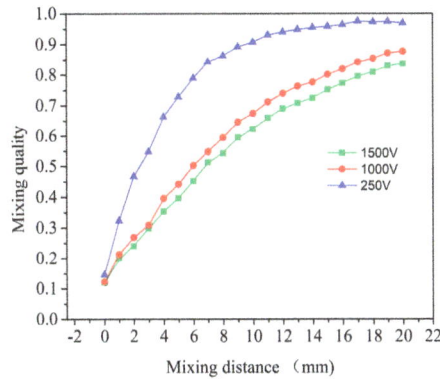

Figure 7. Comparison of mixing quality in Electro-osmotic flow (EOF) along the main channel with three electric potential: 250 V, 1000 V, 1500 V.

4.3. Comparison of Two Modes

In this work, we aim to ellaborate the difference between these two driven modes and investigate the effect of flow field on mixing in both modes. In this section, the driven methods are compared in terms of flow behavior and mixing quality.

Fisrt, we compare the flow field distribution in these two modes. As shown in Figure 8a, with a same flow rate at 2.5 μL/min, the flow field distribution differs under different driven methods. As discussed above in Sections 4.1 and 4.2, the flow fields differ from typical Poiseuille flow and electroosmotic flow, due to the influence of herringbone pattern. In comparison, in the electro-osmotic flow, the velocity shows a much lower gradient in cross direction.

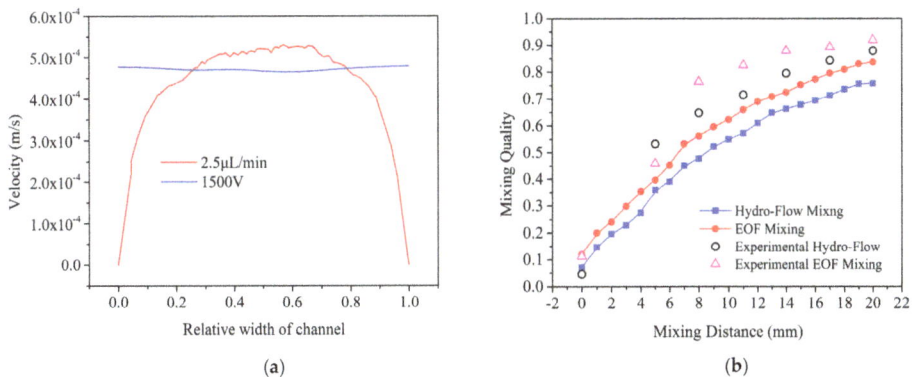

(a)

(b)

Figure 8. (a) Comparison of flow field distribution under two driven modes, the flow rate at outlet equals in both modes; (b) Comparison of mixing quality under both driven modes, and comparison simulation results and experimental results.

Figure 8b compares the mixing quality in both driven methods. Numerical results shows that the mixing quality improved about 4–8% under electro-osmotic driven mode compared with that of hydrodynamic flow, which is strictly close to the experimental results [23]. As discussed above, flow field is key to mixing quality. With a same geometry and total flow rate, the main factor inflencing mixing quality is local flow field distrbution. Thus, we can conclude that the main reason leading to the difference of mixing quality under these two driven modes is the distribution of flow field.

5. Conclusions

In this work, we utilize CFD method to investigate the passive mixing process under two driven methods, hydrodynamic flow and electro-osmotic flow, at an extremely low *Re* regime. The mechanisms of fluid and mass transfer processes inside the micro mixer have been studied and compared. The effects of flow rate have been investigated and analyzed theoretically. Based on the simulation, we can conclude that:

1. In the simulation of passive mixing in both pressure driven flow and electroosmotic flow, the mixing quality is improved with a lower flow rate. Lower flow rate ensures a longer mean residence time. It is noted that diffusion takes the domineering factor in passive mixing in both method. While increasing the flow rate will facilitate the local diffusion, at extremely low *Re* regime, the longer residence time ensures the promotion of mixing performance.
2. Both in hydrodynamic flow and electro-osmotic flow, flow field is influenced by the herringbone structure. In hydrodynamic flow, the flow field is from typical Poiseuille flow, while in EOF the distribution of flow field is different from typical flat flow distribution in EOF.
3. The reason in the difference of mixing quality in two driven modes is the flow field distribution. With a same total flow rate, flow field distribution is totally different in pressure driven flow and electroosmotic flow. Moreover, at extremely low *Re* regime mixing performance is better in EOF than that in pressure driven flow.

Acknowledgments: This work has been jointly supported by National Natural Science Foundation of China (Grant No. 51376039, Grant No. 21575019) and the Fundamental Research for the Central Universities of China (Grant No. N160302001). Dongyang Wang also gives his great appreciations to China Scholarship Council (Grant No. 201606350174).

Author Contributions: Dongyang Wang and Yang Gao proposed the numerical model and performed the simulation; Ming Hao, Dongyang Wang and Qi Mei analyzed the data; Zhiyong Wu contributed the experimental data; Dongyang Wang wrote the paper, Dechun Ba and Kun Liu supported this work with their funding.

Conflicts of Interest: The authors declare no conflict of interest.

Appendix A

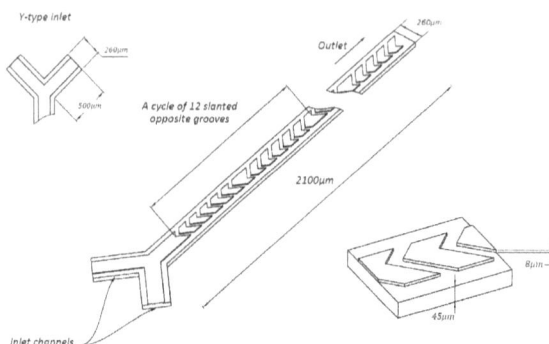

Figure A1. Geometry schematic of the model in micro mixing system and cycled pattern.

The mixing system consists of a pair of Y-type inlets, a main channel and an outlet, all have a rectangular cross-section. Two incoming channels have a width of 260 μm and a length of 500 μm; the width of the main channel is 260 μm and the length is 2100 μm. The height of all the channels is 53 μm. The main channel comprises 158 slanted grooves in, forming 19 repeated herringbone structure cycles. The dimension of the grooves is 8μm in depth. Figure 1 describes the main schematic of the micro mixing system.

Appendix B

The governing equations in simulation of mixing in pressure driven flow and EOF are as follows:
Mixing in pressure driven flow:

$$\nabla\left(\vec{u}\right) \tag{A1}$$

$$\rho(D\vec{u}/Dt) = \rho\left(\partial\vec{u}/\partial t + \vec{u}\cdot\nabla\vec{u}\right) = -\nabla P + \mu\nabla^2\vec{u} + \vec{F} \tag{A2}$$

$$\partial c/\partial t + \left(\vec{u}\cdot\nabla\right)c = d\nabla^2 c. \tag{A3}$$

Mixing in EOF:

$$\nabla\left(\vec{u}\right) \tag{A4}$$

$$\rho(D\vec{u}/Dt) = \rho\left(\partial\vec{u}/\partial t + \vec{u}\cdot\nabla\vec{u}\right) = -\nabla P + \mu\nabla^2\vec{u} + \rho_e\cdot\vec{E} \tag{A5}$$

$$\vec{E} = \nabla\Psi \tag{A6}$$

$$\nabla^2\Psi = -\rho_e/\varepsilon\varepsilon_0 \tag{A7}$$

$$\partial c/\partial t + \left(\vec{u}\cdot\nabla\right)c = d\nabla^2 c. \tag{A8}$$

Appendix C

A preliminary grid dependence has been conducted to minimize the influence of mesh number on the resulting mixing efficiency. The number of the meshes in the mesh independence testing varies from 45,165 to 5,343,381. When the element increases from 1,139,569 to 1,870,067, the resulting mixing quality at outlet shows almost no difference for every simulation input, while the mesh quality increases from 0.7369 to 0.7438. To obtain a more reliable result, the final number of elements of the mesh in this study is 1,870,067, with a 0.7438 mesh quality. Figure A2 displays the mesh grids profile. The free tetrahedral mesh is applied to the whole domain and then calibrated for fluid dynamics. The size of the mesh was customized, varying from 1.6 μm to 4 μm. Mesh modification was then performed, a curvature factor of 0.5 was used with a resolution of narrow regions remaining at 0.8.

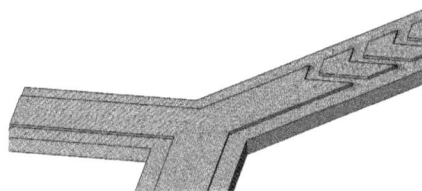

Figure A2. Mesh configuration of the domain at the Y-junction.

References

1. Whitesides, G.M. The origins and the future of microfluidics. *Nature* **2006**, *442*, 368–373. [CrossRef] [PubMed]
2. Jia, Y.; Zhang, Z.X.; Su, C.; Lin, Q. Isothermal titration calorimetry in a polymeric microdevice. *Microfluid. Nanofluid.* **2017**, *21*, 90. [CrossRef]
3. Pagliara, S.; Franze, K.; McClain, C.R.; Wylde, G.; Fisher, C.L.; Franklin, R.J.M.; Kabla, A.J.; Keyser, U.F.; Chalut, K.J. Auxetic nuclei in embryonic stem cells exiting pluripotency. *Nat. Mater.* **2014**, *13*, 638–644. [CrossRef] [PubMed]
4. Petkovic, K.; Metcalfe, G.; Chen, H.; Gao, Y.; Best, M.; Lester, D.; Zhu, Y. Rapid detection of Hendra virus antibodies: An integrated device with nanoparticle assay and chaotic micromixing. *Lab Chip* **2017**, *17*, 169–177. [CrossRef] [PubMed]
5. Sasaki, N.; Kitamori, T.; Kim, H.B. AC electroosmotic micromixer for chemical processing in a microchannel. *Lab Chip* **2006**, *6*, 550–554. [CrossRef] [PubMed]
6. Rhee, M.; Burns, M.A. Microfluidic pneumatic logic circuits and digital pneumatic microprocessors for integrated microfluidic systems. *Lab Chip* **2009**, *9*, 3131–3143. [CrossRef] [PubMed]
7. Xu, Q.B.; Hashimoto, M.; Dang, T.T.; Hoare, T.; Kohane, D.S.; Whitesides, G.M.; Langer, R.; Anderson, D.G. Preparation of Monodisperse Biodegradable Polymer Microparticles Using a Microfluidic Flow-Focusing Device for Controlled Drug Delivery. *Small* **2009**, *5*, 1575–1581. [CrossRef] [PubMed]
8. Liang, T.F.; Zou, X.Y.; Mazzeo, A.D. A Flexible Future for Paper-based Electronics. In Proceedings of the SPIE Defense + Security, Baltimore, MD, USA, 17 May 2016.
9. Nguyen, N.T.; Hejazian, M.; Ooi, C.H.; Kashaninejad, N. Recent Advances and Future Perspectives on Microfluidic Liquid Handling. *Micromachines* **2017**, *8*, 186. [CrossRef]
10. Hessel, V.; Lowe, H.; Schonfeld, F. Micromixers—A review on passive and active mixing principles. *Chem. Eng. Sci.* **2005**, *60*, 2479–2501. [CrossRef]
11. Lee, C.Y.; Wang, W.T.; Liu, C.C.; Fu, L.M. Passive mixers in microfluidic systems: A review. *Chem. Eng. J.* **2016**, *288*, 146–160. [CrossRef]
12. Chun, H.; Kim, H.C.; Chung, T.D. Ultrafast active mixer using polyelectrolytic ion extractor. *Lab Chip* **2008**, *8*, 764–771. [CrossRef] [PubMed]
13. Lu, L.H.; Ryu, K.S.; Liu, C. A magnetic microstirrer and array for microfluidic mixing. *J. Microelectromech. Syst.* **2002**, *11*, 462–469.
14. Yaralioglu, G.G.; Wygant, I.O.; Marentis, T.C.; Khuri-Yakub, B.T. Ultrasonic mixing in microfluidic channels using integrated transducers. *Anal. Chem.* **2004**, *76*, 3694–3698. [CrossRef] [PubMed]
15. Lu, Z.H.; McMahon, J.; Mohamed, H.; Barnard, D.; Shaikh, T.R.; Mannella, C.A.; Wagenknecht, T.; Lu, T.M. Passive microfluidic device for submillisecond mixing. *Sens. Actuator B Chem.* **2010**, *144*, 301–309. [CrossRef] [PubMed]
16. Li, Y.; Zhang, D.L.; Feng, X.J.; Xu, Y.Z.; Liu, B.F. A microsecond microfluidic mixer for characterizing fast biochemical reactions. *Talanta* **2012**, *88*, 175–180. [CrossRef] [PubMed]
17. Feng, X.S.; Fu, Z.; Kaledhonkar, S.; Jia, Y.; Shah, B.; Jin, A.; Liu, Z.; Sun, M.; Chen, B.; Grassucci, R.A.; et al. A Fast and Effective Microfluidic Spraying-Plunging Method for High-Resolution Single-Particle Cryo-EM. *Structure* **2017**, *25*, 663–670. [CrossRef] [PubMed]
18. Stroock, A.D.; Dertinger, S.K.W.; Ajdari, A.; Mezic, I.; Stone, H.A.; Whitesides, G.M. Chaotic mixer for microchannels. *Science* **2002**, *295*, 647–651. [CrossRef] [PubMed]
19. Lehmann, M.; Wallbank, A.M.; Dennis, K.A.; Wufsus, A.R.; Davis, K.M.; Rana, K.; Neeves, K.B. On-chip recalcification of citrated whole blood using a microfluidic herringbone mixer. *Biomicrofluidics* **2015**, *9*. [CrossRef] [PubMed]
20. Sheng, W.A.; Ogunwobi, O.O.; Chen, T.; Zhang, J.L.; George, T.J.; Liu, C.; Fan, Z.H. Capture, release and culture of circulating tumor cells from pancreatic cancer patients using an enhanced mixing chip. *Lab Chip* **2014**, *14*, 89–98. [CrossRef] [PubMed]
21. Toth, E.L.; Holczer, E.G.; Ivan, K.; Furjes, P. Optimized Simulation and Validation of Particle Advection in Asymmetric Staggered Herringbone Type Micromixers. *Micromachines* **2015**, *6*, 136–150. [CrossRef]
22. Aubin, J.; Fletcher, D.F.; Xuereb, C. Design of micromixers using CFD modelling. *Chem. Eng. Sci.* **2005**, *60*, 2503–2516. [CrossRef]

23. Ansari, M.A.; Kim, K.Y. Shape optimization of a micromixer with staggered herringbone groove. *Chem. Eng. Sci.* **2007**, *62*, 6687–6695. [CrossRef]

24. Williams, M.S.; Longmuir, K.J.; Yager, P. A practical guide to the staggered herringbone mixer. *Lab Chip* **2008**, *8*, 1121–1129. [CrossRef] [PubMed]

25. Li, C.A.; Chen, T.N. Simulation and optimization of chaotic micromixer using lattice Boltzmann method. *Sens. Actuator B Chem.* **2005**, *106*, 871–877.

26. Fang, F.; Zhang, N.; Liu, K.; Wu, Z.Y. Hydrodynamic and electrodynamic flow mixing in a novel total glass chip mixer with streamline herringbone pattern. *Microfluid. Nanofluid.* **2015**, *18*, 887–895. [CrossRef]

micromachines

MDPI

Article

Performance Analysis and Numerical Evaluation of Mixing in 3-D T-Shape Passive Micromixers

Mahmut Burak Okuducu [1] and Mustafa M. Aral [2,*]

[1]   School of Civil and Environmental Engineering, Georgia Institute of Technology, Atlanta, GA 30332, USA;
mbokuducu@gatech.edu

[2]   Department of Civil Engineering, Bartin University, Bartin 74100, Turkey

*   Correspondence: mustafaaral@bartin.edu.tr; Tel.: +90-738-223-5036

Received: 2 April 2018; Accepted: 25 April 2018; Published: 28 April 2018

Abstract: In micromixer devices, laminar characteristics of the flow domain and small diffusion constants of the fluid samples that are mixed characterize the mixing process. The advection dominant flow and transport processes that develop in these devices not only create significant challenges for numerical solution of the problem, but they are also the source of numerical errors which may lead to confusing performance evaluations that are reported in the literature. In this study, the finite volume method (FVM) and finite element method (FEM) are used to characterize these errors and critical issues in numerical performance evaluations are highlighted. In this study, we used numerical methods to evaluate the mixing characteristics of a typical T-shape passive micromixer for several flow and transport parameters using both FEM and FVM, although the numerical procedures described are also equally applicable to other geometric designs as well. The outcome of the study shows that the type of stabilization technique used in FEM is very important and should be documented and reported. Otherwise, erroneous mixing performance may be reported since the added artificial diffusion may significantly affect the mixing performance in the device. Similarly, when FVM methods are used, numerical diffusion errors may become important for certain unstructured discretization techniques that are used in the idealization of the solution domain. This point needs to be also analyzed and reported when FVM is used in performance evaluation of micromixer devices. The focus of this study is not on improving the mixing performance of micromixers. Instead, we highlight the bench scale characteristics of the solutions and the mixing evaluation procedures used when FVM and FEM are employed.

Keywords: micromixers; microfluidics; CFD; grid type; finite volume method; finite element method; numerical diffusion; artificial diffusion; false diffusion

1. Introduction

During the past two decades, numerous experimental and numerical modeling studies have been performed for the design of microfluidic devices in which various geometric designs are recommended to improve the mixing efficiency in these devices. Due to widespread use of microfluidic devices in diverse disciplines it is obvious that these studies will grow exponentially in the future with the main goal of improving the mixing performance. Micromixers are essential components of microfluidic systems [1] in which fluid mixing is developed in miniaturized devices. These devices are extensively used in chemical, biological, medical, and environmental applications [2,3]. Fast and thorough mixing of two or more samples at micro scales are the purpose of micromixers. Major benefits of microfluidic devices, as opposed to their macro counterparts, are their high surface-to-volume ratios, small amount of sample requirement for analysis, low cost, short time operating conditions, high efficiency, compactness of the device, and safety in case of hazardous chemical usage [2,4]. Micromixers are usually categorized as active and passive devices depending on the presence of external disturbance effects which are employed to enhance mixing efficiency in active micromixers [2,5–7]. Due to important advantages of passive micromixers [8,9], passive micromixers have been widely used and investigated by researchers in which improved flow, transport and mixing behavior of these devices are studied.

In macro-scale systems, mixing mainly develops due to turbulent flow characteristics by which liquids are naturally stirred and the impact of molecular diffusivity in turbulent flow regimes can be ignored since turbulent diffusivity is the dominant mixing parameter [10]. Mixing at micro-scales is a more challenging process due to strictly laminar flow regimes that are observed in these devices where mixing is performed at very low molecular diffusivities (e.g., $D = 10^{-9}-10^{-11}$ m^2/s [11]) for many chemical and biological solutions. The advection dominant transport regimes (i.e., high Peclet (Pe) number) that develop in these devices lead to problems in numerical solution of these problems.

Computational Fluid Dynamics (CFD) simulations are usually employed to investigate fluid flow and solute transport at micro scales. The finite volume method (FVM) and finite element method (FEM) are the two numerical techniques that are commonly used in these applications. Complications in numerical solution often arise depending on the characteristics of the numerical method employed along with other factors such as discretization techniques, grid size and properties, and boundary conditions used. In aggregate, these choices may significantly affect the reliability of the numerical performance evaluation results. The two-important numerical complication in CFD studies are the numerical diffusion (or false diffusion) and numerical dispersion effects. These effects may prevent the proper interpretation of the results of fluid flow and mixing analysis performed. Numerical diffusion arises from the numerical approximation of the advection term in the flow equation [12–14]. On the other hand, lack of dispersion yields numerical instability problems (i.e., oscillations) in the solution because of high gradients of the variables that appear in the flow and transport domain. These oscillations need to be eliminated as much as possible to obtain reliable numerical solutions. The degree of numerical errors in these solutions are associated with the discretization schemes that are used in the solution of the governing equations (i.e., the magnitude of truncation errors) as well as other numerical and physical parameters used, such as grid size, grid type, and diffusivity (or viscosity for fluid flow). Godunov [15] showed that a linear monotone scheme which does not create over- and undershoots can be at most first-order accurate. However, higher order schemes are more commonly preferred in applications due to their lower numerical diffusion effects as it is preferred in this study as well. In CFD applications, especially for advection dominant problems, FVM seems to be more advantageous and may provide relatively more consistent results [16] because of its conservative mass, momentum, and energy formulation structure. It is also shown that if the flow direction is orthogonal to the grid lines or in other words, the flow direction and grid lines are orthogonally aligned in the computational domain, the FVM does not produce high false diffusion effects [17]. The amount of false diffusion created increases when the angle between streamlines and gridlines approaches to 45°. In most CFD applications, maintaining mesh and flow alignment in a computational domain in this sense is not practical and numerical solution inevitably exhibits some effect of false diffusion which needs to be accounted for in mixing performance evaluation. Meanwhile, FEM suffers from stability as well as numerical diffusion problems, especially when working with advection dominant systems [18]. In this application, although it is possible to avoid unwanted node-to-node oscillations by grid refinements, this approach is not practical because of high cost of numerical simulations at fine grids that are necessary. A practical approach to stabilize oscillations in FEM is known as artificial diffusion (or artificial viscosity for fluid flow) [17,19] in which diffusion constant (or viscosity) is artificially increased to eliminate the instability in the numerical solution at the cost of excess diffusion that is added to the system. This approach may affect the interpretation of the outcome of the mixing performance of micromixers since the excess diffusion changes the physics of the problem and for such cases it is not clear if mixing outcome is artificial due to the added artificial diffusion and false diffusion effects or real due to the molecular diffusion of the fluids that are mixed. Researchers have introduced several techniques for both FVM and FEM to solve fluid flow and mass transport equations more accurately by suppressing the negative effects of false diffusion and artificial diffusion in CFD applications. Unfortunately, none of these methods are problem-free. The most popular techniques used in stabilization of FEM and high-resolution schemes applied in FVM can be found in [14,20], respectively, with evidence of deficiencies in both of these techniques.

Significant effort is necessary in reporting the outcome of the mixing performance in numerical studies. As stated earlier, in micromixers molecular diffusion effect is typically the only mixing mechanism. The mixing effects of diffusion may be acutely masked by the aforementioned numerical errors. Since complete elimination of numerical errors is not possible, quantifying the presence of these errors are necessary to provide reliable and unsusceptible results in the evaluation of mixing performance of micromixers.

In investigating several micromixer studies, Liu [13] extensively discussed the extent of numerical diffusion problems in micromixer literature. He also stated that numerical diffusion effects of high-order schemes were rarely discussed in the literature, with which we agree. In his study, Liu conducted several tests using different numerical schemes under various grid size, flow, and transport conditions for a three-dimensional microchannel mixer. He utilized hexahedral mesh type in FVM and proposed a set of equations to quantify average numerical diffusion which results from both flow and scalar transport solutions. He also studied tetrahedral mesh type and advised that the use of this type of a mesh should be avoided especially for scalar transport solution since the amount of false diffusion and computational cost is higher than that of hexahedral mesh type. By examining various higher order schemes (e.g., Second-Order Upwind, QUICK [21], MUSCL [22]), he found that false diffusion can be significantly reduced if higher order schemes are used instead of first order upwind schemes.

Bailey [12] investigated false diffusion effects using FVM in a two-dimensional test problem for both structured and unstructured grids applying first and second-order upwind methods and quantified the amount of false diffusion that is generated. He also developed a set of procedures to estimate and manage the required grid size by which the false diffusion amount can be reduced in steady micro scale mixing simulations.

Soleymani et al. [23] investigated flow dynamics and mixing in a T-shape micromixer using FVM and discussed several approaches to study and diminish numerical diffusion effects at moderate cell Pe numbers. They performed spatial discretization using high-order QUICK scheme, locally refined the mesh, and increased the diffusion constant artificially to overcome high computation cost. In their study they reported that numerical simulation results were used for optimization purposes rather than the physical mixing evaluation of the micromixer.

On the other hand, in several micromixer studies [8,24–28] where FEM was employed, the authors did not report the instability and false diffusion effects although they have studied advection dominant systems. In these applications, depending on the stabilization method applied, mixing performance results reported may show significant variations depending on the degree of numerical treatment applied.

In view of these arguments, the purpose of this study is to systematically investigate, quantify, and show the numerical diffusion effects in micromixer CFD simulations. For this purpose, various fluid flow, transport, and idealizations (grid configurations) were utilized when both FVM and FEM is used in solution. Quantification and recommendation of application ranges of these techniques are provided for T-shape mixers for future reference. However, the analytical methods used in the evaluation are generic and applicable to other mixer devices as well. Although both FVM and FEM is used in this study, the main emphasis is on FVM due to its extensive usage in micromixer applications and widespread availability of this method in several commercial and non-commercial CFD packages.

2. Mathematical Model

In micro channels, fluid flow and transport equations are defined using a set of partial differential equations which are derived from the well-known mass, momentum, and energy conservation principles. Here we will assume that mixing fluids are of constant density, constant viscosity, miscible, and non-reactive with identical physical properties. Gravitational and temperature effects were not considered in the mathematical model due to their negligible effects in micromixer applications. The flow regime is assumed to be laminar and the fluid is incompressible. Based on these assumptions, the governing equations which describe the flow field and mass transport are the continuity equation, Equation (1); Navier–Stokes equation, Equation (2); and the advection-diffusion (AD) equation, Equation (3).

$$\nabla \cdot \mathbf{u} = 0 \tag{1}$$

$$\rho\left[\frac{\partial \mathbf{u}}{\partial t} + \mathbf{u} \cdot \nabla \mathbf{u}\right] = -\nabla p + \mu \nabla^2 \mathbf{u} \tag{2}$$

$$\frac{\partial c}{\partial t} + \mathbf{u} \cdot \nabla c = D \nabla^2 c \tag{3}$$

Equations (1) and (2) represent mass and momentum conservation respectively, where, ρ is the density of fluid (kg/m^3), μ is the dynamic viscosity of the fluid (Pa·s), $\mathbf{u}$ is the velocity vector (m/s), p is the pressure (Pa). For all scenarios, the stationary velocity field obtained from the solution of Equations (1) and (2) is used to simulate the steady-state transport of solute using Equation (3) in which c is the concentration of the solute (mol/m^3) and D is the molecular diffusion coefficient (m^2/s). Throughout this study, the molecular diffusion coefficient of the transported solute is chosen as $D = 3 \times 10^{-10}$ m^2/s which corresponds to crystal violet dye [29].

In this study, flow and mixing analysis of two liquids were investigated in a typical three-dimensional (3-D) T-shape passive micromixer which consists of two equal length inlet channels and a mixing channel as shown in Figure 1. Although the analytical techniques used for mixing evaluation is applicable to other micromixer geometries, in this study our focus will be on T-mixers. Inlet channel lengths were chosen as 500 μm with a square cross section of 100 μm × 100 μm considering that flow will be fully developed before entering the mixing channel for the highest Reynolds number (Re) scenario tested in this study (i.e., Re = 100). The mixing channel length is 1000 μm with a width (W) and height (H) of 200 μm and 100 μm respectively. The T-shape passive micromixer dimensions are consistent with the T-shape passive micromixers studied in the literature [25,30,31].

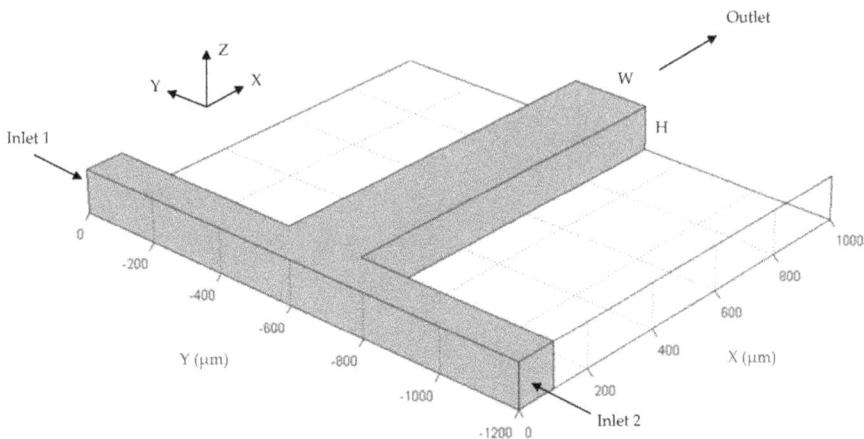

Figure 1. Three-dimensional view of T-shape passive micromixer.

Physical properties of mixing fluids and appropriate boundary conditions used in simulations are given in Table 1.

Table 1. Fluid properties and boundary conditions.

Simulation Type	Material Properties	Boundary	Boundary Condition
Incompressible Fluid Flow	$\rho = 1000$ kg/m^3 $\mu = 0.001$ Pa·s	Inlet 1 Inlet 2 Outlet Walls	Uniform Inflow Uniform Inflow $p = 0$ No-Slip
Advective-Diffusive Transport	$D = 3 \times 10^{-10}$ m^2/s	Inlet 1 Inlet 2 Outlet Walls	$c = 0$ mol/m^3 $c = 1$ mol/m^3 $\partial c/\partial \mathbf{n} = 0$ * $\partial c/\partial \mathbf{n} = 0$

* **n** is the unit vector outer normal to the wall.

In micromixer studies, flow and transport characteristics can be defined through two dimensionless numbers, i.e., the Reynolds (Re) number and Peclet (Pe) number, as given in Equation (4), respectively,

$$\text{Re} = \frac{\overline{U}D_h}{v}; \quad \text{Pe} = \frac{\overline{U}D_h}{D} \tag{4}$$

where, $\overline{U}$ is the average flow velocity (m/s), v is the kinematic viscosity (m^2/s), D_h is the characteristic length which is assumed to be hydraulic diameter of a rectangular duct (m) and defined as, $D_h = 4A/P$ in which A is the area and P is the wetted perimeter of the duct.

Typically, Re = 2300 is considered as the critical point [2,32] after which flow regime starts to change from laminar flow to turbulent flow. In microfluidic applications, Re number is generally far below than this transition point (usually Re < 100 [3]). Therefore, the flow regime in micro channels can be evaluated as strictly laminar. High Peclet number indicates that mass transport is due to advection rather than diffusion. In this study, Re and Pe numbers were calculated in the mixing channel of T-mixer. In addition to these characteristic numbers, the cell Peclet number (Pe$_\Delta = U\Delta x/D$) and cell Reynolds number (Re$_\Delta = U\Delta x/v$) are used to observe the accuracy and stability of the numerical solution. For example, in case of scalar transport, when cell Peclet number is greater than 1, FEM yields oscillatory solutions since large concentration gradients could not be adequately captured. Similarly, several second or higher-order numerical schemes used in FVM (e.g., central difference, second-order upwind etc.), show numerical instability when cell Peclet number is greater than 1. In this analysis, cell Peclet and cell Reynolds numbers were calculated as an average value since the velocity magnitude is a local variable in the computational domain.

The mixing performance of simulations were quantified using mixing index (MI) as shown in Equation (5) [32,33] in which σ is the standard deviation of the concentration in a given cross-section, c_i and c_m are concentrations at ith sample point and mean concentration on the cross-section respectively, and N is the total number of sample points.

$$\text{MI} = 1 - \sigma; \quad \sigma = \sqrt{\frac{1}{N}\sum_{i=1}^{N}\left(\frac{c_i - c_m}{c_m}\right)} \tag{5}$$

Mixing efficiency changes between 0 and 1 corresponding to unmixed (i.e., 0% mixing) and completely mixed (i.e., 100% mixing) states, respectively. An acceptable mixing efficiency in micromixer applications is usually higher than 80% [32].

The average numerical diffusion in the solution domain can be quantified using the procedure proposed in [13]. In this approach, effective diffusivity in a numerical solution is defined as the sum of the physical diffusivity in the system and the numerical diffusion as given in Equation (6). Here, D_M is molecular diffusivity of the mixing fluid and D_N is numerical or false diffusivity.

$$D_{Effective} \approx D_M + D_N \tag{6}$$

The effective diffusivity given in Equation (6) can be computed using the results of numerical solution of scalar transport equation as formulated in Equation (7).

$$D_{Effective} = \frac{c_{inlet}^2 - c_{outlet}^2}{2\tau(\nabla c)_\forall^2} \tag{7}$$

where,

$$c_{inlet}^2 = \frac{1}{Q} \int_A \mathbf{n} \cdot \mathbf{u} c^2 dA \tag{8}$$

$$c_{outlet}^2 = \frac{1}{Q} \int_A \mathbf{n} \cdot \mathbf{u} c^2 dA \tag{9}$$

$$\nabla c_\forall^2 = \frac{1}{\forall} \int_\forall (\nabla c) \cdot (\nabla c) d\forall \tag{10}$$

In these equations, c is concentration, $\mathbf{n}$ is unit vector, $\forall$ is the volume of the micromixer, Q is the volumetric flow rate, c_{inlet}^2 and c_{outlet}^2 are flow rate weighted average value of concentration at the inlet and outlet respectively, $\nabla c_\forall^2$ is the volume average, τ is the mean residence time of the flow, $\tau = \forall/Q$ for unobstructed T-shape mixers. For obstructed micromixers, actual residence time of solute particles must be used in Equation (7). Given Equation (7), numerical diffusivity can be calculated by setting the molecular diffusion constant to zero and the effective diffusivity calculated would give the average value of numerical diffusion expected in the solution of the scalar transport equation. Therefore, in this study, two simulations were performed for each scalar transport scenario. In the first simulation, molecular diffusion constant was set to zero and the AD equation was solved and from the numerical solution the effective diffusivity was calculated using Equation (7) which in this case would equal to the estimated numerical diffusion in the solution ($D_{Effective} \approx D_N$). The second simulation was conducted using the molecular diffusion constant of $D_M = 3 \times 10^{-10}$ m^2/s. For this case the effective diffusivity calculated in Equation (6) will now include the effects of both the molecular diffusion and numerical diffusion as described in Equation (6) ($D_{Effective} \approx D_M + D_N$). The analysis of the results of these two solutions will be discussed in the following sections.

In this study, numerical solution of the governing partial differential equations was carried out using both FVM and FEM with two different CFD tools. These are OpenFOAM (v5.0, OpenFOAM Foundation, OpenCFD Ltd., Bracknell, UK) and COMSOL Multiphysics (v5.3a, COMSOL AB, Stockholm, Sweden). OpenFOAM is a non-commercial and open source CFD software based on FVM application. In FVM both structured and unstructured meshes can be utilized. To solve the governing equations for steady-state, incompressible, and laminar flow *simpleFOAM* solver was used in which the SIMPLEC (semi-implicit method for pressure linked equations-consistent) [34] algorithm was used for solving pressure-velocity coupling. Advection and diffusion terms in the flow equation were discretized using second-order upwind scheme (i.e., Gauss linearUpwind) and second-order central difference scheme (i.e., Gauss linear) respectively. The AD equation was solved utilizing the *scalarTransportFoam* solver in which advection terms were discretized using first-order upwind and multiple second-order accurate numerical schemes built-in OpenFOAM (e.g., second-order upwind [35], QUICK, MUSCL, *limitedLinear* [36]) to observe and ensure the boundedness of concentrations for both structured and unstructured meshes used in this study. Diffusion terms in AD equation was treated using second-order central difference scheme. In all simulations, iterations were continued until final residuals of flow and transport equations fall below 1×10^{-12} which was assumed to yield converged solutions. The structured and unstructured meshes used in OpenFOAM simulations were generated using Gmsh [37].

COMSOL Multiphysics, which is a FEM based commercial CFD package, was employed to analyze the artificial diffusion effects. Fluid flow and scalar transport were simulated using the laminar flow and transport of diluted species tools and computational mesh was generated using the geometry interface in the software. For uniformity, discretization accuracy, and convergence levels of flow and transport equations were set equivalent to that of OpenFOAM simulations which was determined from grid convergence analysis. To dampen the oscillation effects in the solution of Navier–Stokes and AD equations, COMSOL provides two stabilization options which are referred to as consistent and inconsistent methods. In the consistent method, directions and gradients of variables are evaluated by the solver and appropriate corrections are made in the regions where stabilization is required. In inconsistent stabilization, physical diffusivity (mass or molecular) is artificially increased with a tuning parameter to obtain a stable solution. In this analysis, simulations were conducted for both correction methods.

FEM and FVM simulations conducted in this study were performed on a personal computer with an Intel Core i7-6900K processor which was overclocked to run at 4.2 GHz (Intel Corporation, Santa Clara, CA, USA) and 64 GB 3200 MHz random-access memory (RAM).

3. Grid Independence Analysis

The numerical error generated in 3-D micromixer analysis is a function of grid size, local velocity magnitude, and orientation of velocity vectors with respect to the grid boundaries which may be identified as grid alignment effects [13]. Grid element type chosen becomes more important in FVM since the non-orthogonal alignment of mesh boundaries and flow direction will increase the effect of false diffusion in the solution since in FVM continuity of the gradients of the unknown parameters across boundaries are used to develop the governing matrix equations. Considering these factors several test cases were designed in this study to observe and quantify the effects of numerical errors. Micromixer simulations were designed for five different flow scenarios (i.e., Re = 0.1, 1, 10, 50, 100) and for three type of mesh structures (i.e., structured hexahedral, structured prism, and unstructured tetrahedral), which are generated using the elements shown in Figure 2. Designed test cases with simulation parameters are given in Table 2.

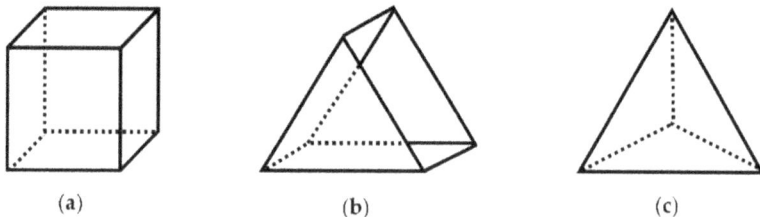

(a)   (b)   (c)

Figure 2. Three-dimensional mesh element types: (a) hexahedron; (b) prism; (c) tetrahedron.

Table 2. Test cases and simulation parameters for hexahedral, prism, and tetrahedral mesh.

Inlet Velocity (m/s)	Mixing Channel	
	Re	Pe
0.00075	0.1	3.33×10^2
0.0075	1	3.33×10^3
0.075	10	3.33×10^4
0.375	50	1.67×10^5
0.75	100	3.33×10^5

Table 2. *Cont.*

Mesh Level	Constant Flow, Re = 100	
	Average Cell Re *	Average Cell Pe *
L_1: Δx = 2.0 μm	1.50	5000
L_2: Δx = 3.0 μm	2.25	7500
L_3: Δx = 4.5 μm	3.38	11250
L_4: Δx = 6.6 μm	4.95	16500
Re	Constant Grid Level, L_1	
	Average Cell Re	Average Cell Pe
0.1	0.0015	5
1	0.015	50
10	0.15	500
50	0.75	2500
100	1.5	5000

* Average cell Re—Pe numbers are calculated for structured hexahedral mesh.

In the T-mixer, the fluid flow solutions showed that if the flow profile in the inlet channels are not fully developed before entering the mixing channel, stratified (or separated) flow regions occur at the beginning of the mixing section implying that the liquids which approach to the mixing channel from two different inlet streams travel side-by-side along the mixing channel without rotation in the z-direction for all Re scenarios. If, however, the flow in the inlet channels are fully developed, periodic (or vortex) flow type may be observed at the entrance section of the mixing channel for Re = 50 and 100 scenarios. Therefore, inlet channel lengths of the T-mixer were selected to be long enough to create a fully developed flow profile for high Re number scenarios studied which is 500 μm. The flow regimes observed in the T-mixer are shown in Figure 3 for Re = 0.1 and 100 cases. In addition, velocity profiles at different cross-sections in the mixing channel are shown in Figure 4. As shown in Figure 4, the most non-uniform flow in the T-mixer occurs at *x* = 200 μm in the mixing channel at Re = 100. Investigation of the fully developed flow effects in the inlet channels on mixing performance is beyond the scope of this study and this point will not be further discussed here. We note that fluid flow results reported in this study agree with the findings of extensive studies reported in the literature on this subject [23,29].

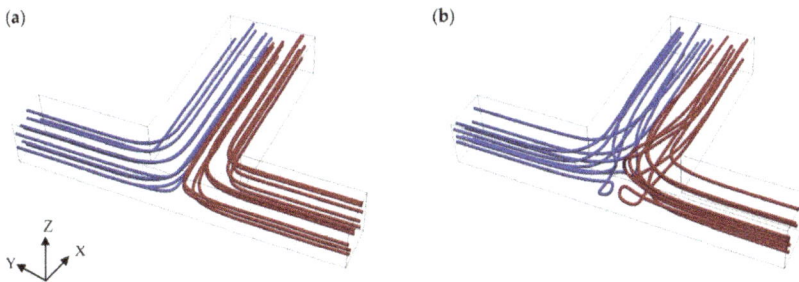

Figure 3. Flow profile at the beginning of mixing channel (Figures show the T-mixer region between *x* = 0 and 500 μm and *y* = −250 and −700 μm): (**a**) separated (or stratified) flow at Re = 0.1; (**b**) periodic (or Vortex) flow at Re = 100.

(a)

x = 100 μm

x = 200 μm

x = 500 μm

x = 1000 μm

(b)

x = 100 μm

x = 200 μm

x = 500 μm

x = 1000 μm

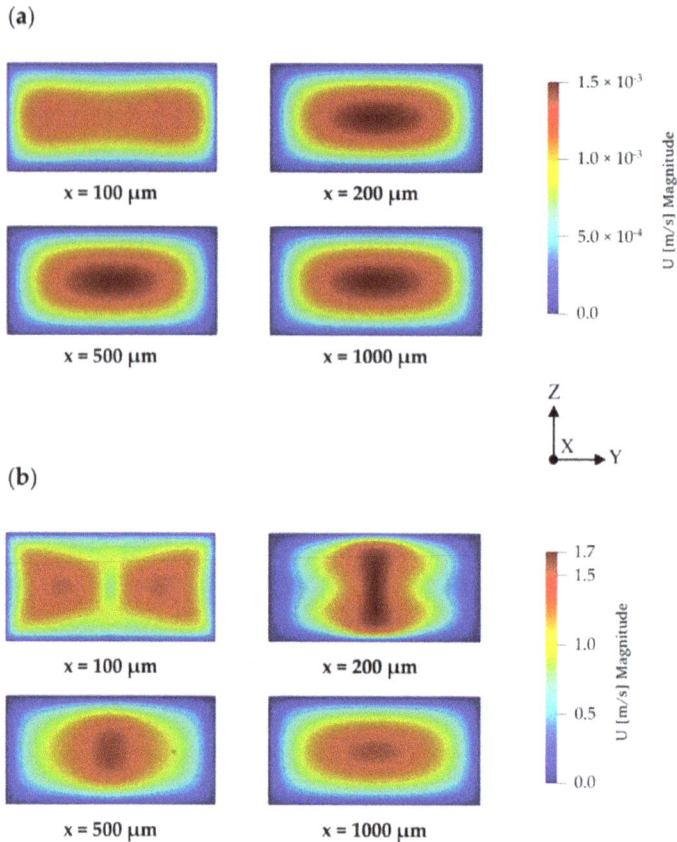

Figure 4. Velocity profile at four cross-sections in the mixing channel of the T-mixer (Color ranges were set to that of $x = 200$ μm plane): (**a**) Re = 0.1; (**b**) Re = 100.

In CFD applications, grid generation is a pivotal stage since spatial discretization errors will inherently affect the numerical solution. Theoretically, it is known that temporal and spatial discretization errors asymptotically approach zero by reducing the time step and mesh element size; however, as the mesh is refined, computational cost and the solution time increases in parallel. Therefore, an optimized solution in terms of accuracy, computational cost, and solution time becomes important. The aim of systematic grid convergence studies in CFD applications is to estimate the amount of spatial discretization error between different mesh densities and determine a practical mesh level at which computational cost is feasible and spatial error is controlled [38]. Grid Convergence Index (GCI) analysis is a well-established and commonly used method for systematic grid convergence studies as described in [39–41]. In this study, GCI method was used to determine the order of convergence of the simulations and verify that solution is independent of mesh resolution. GCI study was made using FVM for three different mesh structures (i.e., structured hexahedral, structured prism, and unstructured tetrahedral) separately for the highest Reynolds number scenario (i.e., Re = 100) since the most complex flow regime was observed at this Re number as can be seen in Figure 3. For hexahedral and prism element types, four different grid levels (i.e., L_1, L_2, L_3, L_4) were created with a constant refinement ratio of 1.5 in x, y, and z coordinate directions. For the tetrahedral grid type, total element number was fixed around the mesh density of prism type as given in Table 3.

Table 3. Mesh properties and Grid Convergence Index (GCI) results for Re = 100 case.

Mesh Level (L)	Grid Size, Δx (µm)	Number of Cells in Computational Domain		
		Hexahedral	Prism	Tetrahedral
L_1	2.0	3.75×10^6	8.70×10^6	8.70×10^6
L_2	3.0	1.09×10^6	2.57×10^6	2.57×10^6
L_3	4.5	3.22×10^5	7.72×10^5	7.74×10^5
L_4	6.6	1.02×10^5	2.46×10^5	2.44×10^5
Mesh Level (L)	Grid Size, Δx (µm)	Max Velocity at x = 200 µm Plane (m/s)		
		Hexahedral	Prism	Tetrahedral
L_1	2.0	1.65617	1.65691	1.65900
L_2	3.0	1.65242	1.65175	1.65398
L_3	4.5	1.64624	1.64361	1.64049
L_4	6.6	1.63322	1.62657	1.61615
Mesh Level Comparison		GCI Between Mesh Levels (%)		
		Hexahedral	Prism	Tetrahedral
L_1–L_2		0.44	0.68	0.22
L_2–L_3		0.57	0.82	0.94
L_3–L_4		0.89	1.18	2.31

GCI study was performed using the flow solution in the T-mixer. Maximum velocity at x = 200 µm plane in the mixing channel was selected as GCI parameter since the most complex flow profile was observed at this point as can be seen in Figure 4b. Crosswise velocity distributions on this plane for three different mesh structures at different mesh densities are shown in Figure 5. It can be seen in Figures 5 and 6, that the GCI study results show that flow field at Re = 100 was resolved consistently for all mesh types using second-order discretization schemes for advection and diffusion terms in Equation (2). Computed velocity values, which are used in GCI study, are approaching to a constant point asymptotically when meshes are refined as plotted in Figure 6a. This asymptotic level can be considered to be the exact solution of flow equations. Similarly, the difference in computed velocities and GCI of two consecutive mesh levels are reduced with grid refinement, as shown in Figure 6b,c.

As given in Table 2, the minimum and maximum average cell Re number scenario (at Re = 100) are 1.50 and 4.95, respectively, for hexahedral mesh. These moderate numbers indicate low advection dominance in the system. In other words, advection and diffusion terms in Equation (2) feed the numerical solution almost equally which reduce numerical instability and false diffusion in the system. This is actually because of the high kinematic viscosity of the fluid. Results showed that, although mesh and flow alignment is not maintained in prism and tetrahedral meshes, the solution was not affected by numerical errors significantly due to resulting moderate cell Re numbers (assuming cell Re number in prism and tetrahedral domain is close to that of hexahedral).

While the velocity profile on the plane at x = 200 µm could be successfully resolved for all mesh types at L_1, as shown in Figure 5d, prism and tetrahedral mesh density is about 2.3 times more than the hexahedral mesh type which is significant in terms of computational cost. The maximum difference in solutions was observed to be 2.31% between L_4 and L_3 of tetrahedral mesh with a mesh density difference of roughly 5×10^5. Also, for this type of idealization, while the difference between solutions at L_2 and L_1 is only 0.22%, L_1 simulations were performed using 6 million more elements. GCI values and mesh density differences between mesh levels are shown in Figure 7.

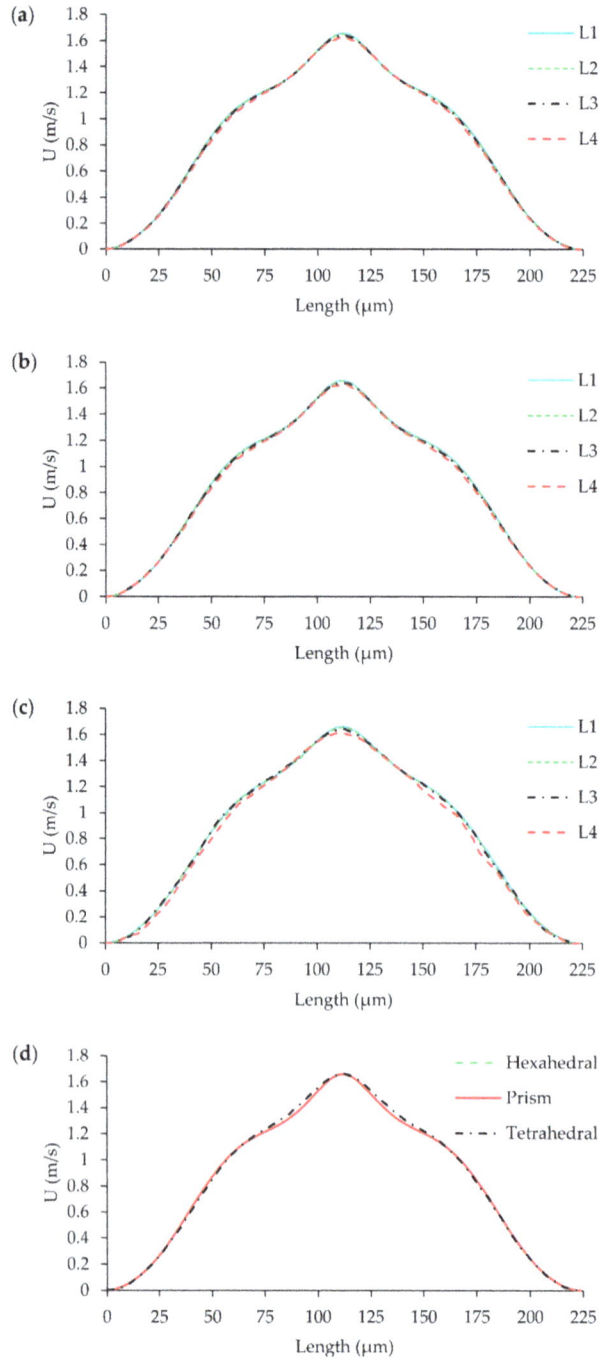

Figure 5. Velocity distribution on the plane at $x = 200$ μm from L_1, L_2, L_3, and L_4 grid level simulations: (**a**) hexahedral; (**b**) prism; (**c**) tetrahedral; (**d**) hexahedral vs. prism vs. tetrahedral solutions at L_1.

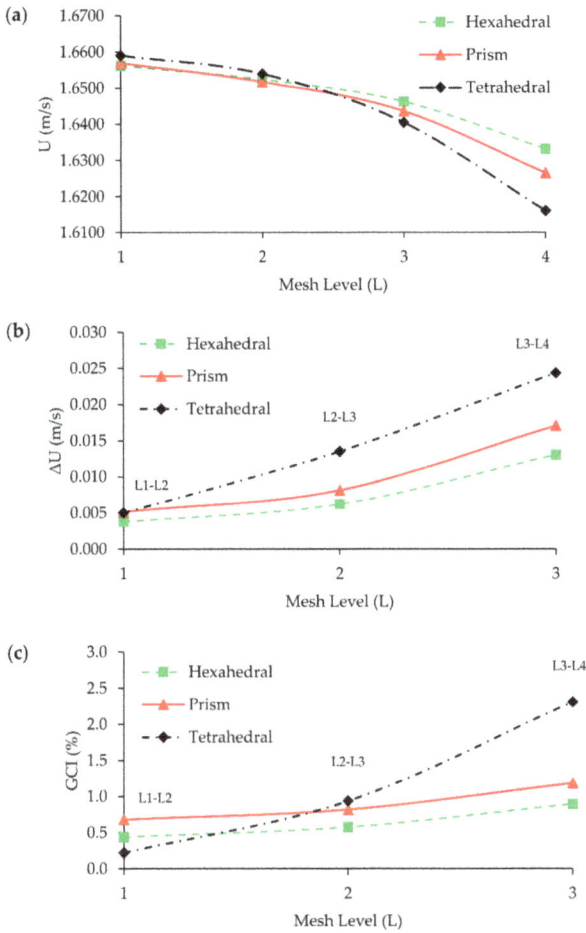

Figure 6. Grid convergence results: (a) maximum velocity magnitude on the plane at $x = 200$ µm obtained from L_1, L_2, L_3, and L_4 grid levels for all mesh types; (b) maximum velocity magnitude difference between mesh levels; (c) GCI between mesh levels.

As it can be seen in Figure 7, sharp changes in the slopes of these lines show that accuracy gain is higher than grid number growth when refinement is maintained between L_4 and L_2 for all grid types. However, after the L_2-L_3 point, mesh density increases excessively with respect to GCI, implying that further grid refinement does not contribute to numerical accuracy significantly, but it increases the computational cost unreasonably. If the 1% GCI difference is determined as the limit, simulations can be done at L_2 for all mesh types, but in this study L_1 mesh density was employed for all test cases and mesh types to investigate the magnitude of numerical diffusion errors in scalar transport solution more accurately. The reason for selection the finest grid for Re = 0.1, 1, 10, 50 scenarios is that each case provides a different average cell Pe number by which the amount of numerical diffusion can be observed. It is clear that increasing the grid size will also increase average cell Pe number. Thus, instead of conducting additional simulations for different grid sizes, all simulations were conducted at the finest grid size and the amount of numerical diffusion was evaluated for various average cell Pe numbers which range between 5 and 5000 for Re = 0.1 and 100 scenarios respectively. These results are provided in Table 2.

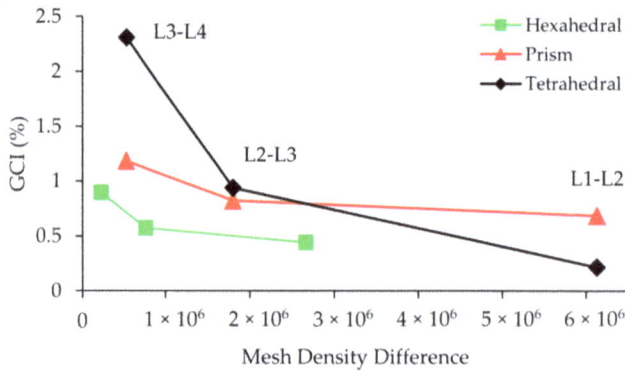

Figure 7. GCI vs. mesh density difference between grid levels.

In the literature, the mixing efficiency parameter is also used in grid refinement analysis. When outlet mixing efficiency is used in grid convergence analysis [9,27,42,43], important discrepancies are observed in grid convergence analysis, especially for prism and tetrahedral mesh types. As it is observed in this study, this discrepancy was also mentioned in [13]. Although mesh refinement reduced numerical errors in the scalar transport solution as shown in Figure 8, mixing efficiencies obtained significantly differ from one another for different mesh types. This is due to numerical errors and numerical diffusion generated during the solution of transport equation which significantly effects the mixing estimates obtained. While hexahedral mesh type predicts outlet mixing efficiency with a 2.33% difference between L_2 and L_1, this amount sharply increases to a value around 9% and 14% at same mesh levels for prism and tetrahedral mesh types, respectively, indicating higher numerical errors and thus higher mixing efficiency.

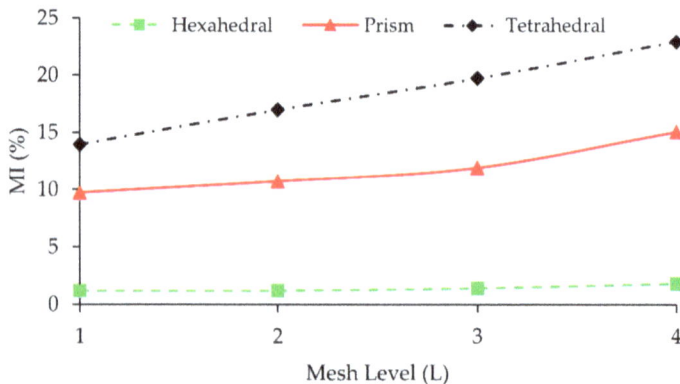

Figure 8. Outlet mixing index of hexahedral, prism, and tetrahedral mesh types at L_1, L_2, L_3, and L_4.

As shown in Figure 9, there is a significant divergence between outlet concentration distributions of different mesh types even at L_1. It needs to be pointed out that use of the mixing efficiency parameter for grid convergence studies may not be appropriate unless cell Pe number is quite low. At high cell Pe numbers, several factors may affect the solution in addition to the grid size and grid type, such as numerical diffusion, oscillations, characteristic of numerical scheme employed, and mesh type etc. In this grid study, the average cell Pe numbers are in the range of 5000 at L_1 and 16,500 at L_4 for the hexahedral mesh type which are significantly high values which require special discretization techniques. This in turn affects the mixing efficiency estimated at the outcome. Even when mesh and flow alignment were maintained in the hexahedral domain, there was a significant difference between mixing index values at the different mesh levels. For moderate cell Pe numbers this problem may be tolerable. However, for high cell Pe number scenarios, the numerical results may imbed important effects of numerical diffusion and thus erroneous mixing efficiency values.

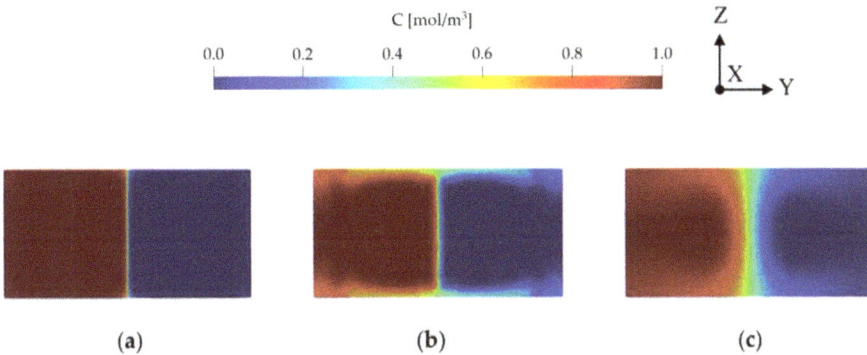

Figure 9. Concentration distribution at the outlet of T-mixer for Re = 100 scenario at L_1: (**a**) hexahedral; (**b**) prism; (**c**) tetrahedral.

The comparison of Figures 7 and 10 shows a significant inconsistency when flow and scalar transport results are considered as grid study parameters. For instance, at Re = 100, scenario difference between L_4 and L_3 was observed as 2.3% for the tetrahedral type when flow field solution is used in GCI, yet this difference increased to more than 20% when the grid study was conducted with a scalar transport solution and mixing efficiency. Even in the best-case scenario, which is the hexahedral mesh solution, there is a considerable disagreement between the results when two grid convergence parameters are compared. The GCI value between L_2 and L_1 is 0.44% when the maximum velocity magnitude is employed in the grid study, whereas, the difference between the same levels increases to 2.33% if outlet mixing efficiency is used in convergence analysis. For other mesh types, this situation is much worse, namely while prism mesh shows around 10% difference between L_2 and L_1, for the tetrahedral mesh type this number is more 20% since the maximum mesh flow disorientation exists in this mesh domain. Based on these observations, the use of the mixing efficiency parameter in grid convergence analysis needs to be approached cautiously. In several micromixer studies [9,27,42,43], selected mesh densities for grid independence analysis are quite close to each other which can cause misinterpretation of grid study outcome especially for high Pe situations. Namely, although grid convergence results may show small differences between mesh densities, the overall effects of numerical diffusion will remain in the simulations of selected mesh density. Therefore, based on the results reported in this study, use of a constant refinement ratio between mesh levels is recommended. Since this approach reveals the discrepancy between results in terms of numerical diffusion effects and mixing efficiency, its use needs to be avoided in GCI studies.

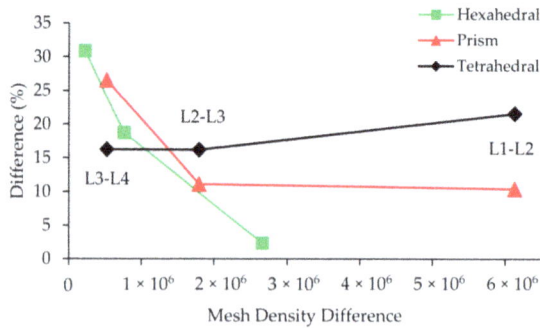

Figure 10. Error (%) between mesh levels in terms of outlet mixing efficiency parameter vs. mesh density difference between grid levels.

4. Results and Discussion

4.1. FVM Analysis

In advection dominant transport, numerical diffusion mainly occurs due to the truncation errors of the advection term. To minimize the amount of numerical diffusion generated several high-order numerical schemes are proposed. Unfortunately, a problem-free solution that overcomes this numerical difficulty entirely is not possible. Although higher order algorithms provide more accurate solutions and may resolve higher gradients, they usually suffer from numerical instabilities which affects the reliability of the numerical solution. In three-dimensional flow systems, higher order schemes need to be used to obtain numerically less diffusive results. In FVM analysis, first-order upwind scheme and several second-order accurate numerical schemes (i.e., QUICK, MUSCL, second-order upwind, and *limitedLinear*) were tested for discretization of advection terms in the AD equation. Simulations showed that QUICK, MUSCL, and second-order upwind schemes showed oscillatory behaviors in some regions depending on the magnitude of the Reynolds number and grid size of structured prism and unstructured tetrahedral mesh types. On the other hand, although amplitudes were not high as for the other two mesh types, tested schemes also exhibited small concentration fluctuations in the hexahedral mesh simulations which varied with Re and grid size. However, stable solutions were obtained for all mesh types, grid sizes, and Re scenarios using the first-order upwind scheme and the *limitedLinear* scheme, which is a type of total variation diminishing scheme (TVD) [14]. Comparison of *limitedLinear* and first-order upwind schemes for L_1 grid size of all mesh types at Re = 100 is given in Figure 11.

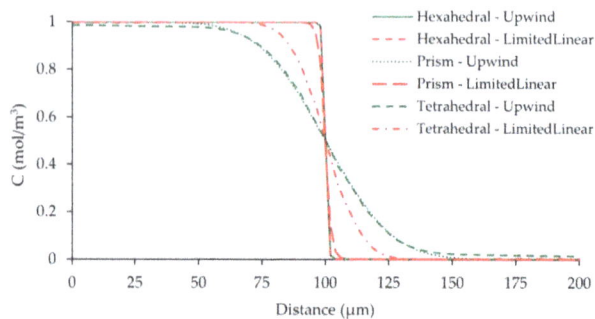

Figure 11. Outlet concentration distribution along the width of the mixing channel at $z = 50$ μm for all mesh types at L_1 grid size at Re = 100 scenario.

Figure 11 shows that the amount of numerical diffusion produced by first-order upwind scheme is very close to that of *limitedLinear* scheme when hexahedral mesh type is used in the computational domain. Therefore, when a good mesh and flow alignment is sustained as existed in hexahedral mesh type, the amount of numerical diffusion produced by first-order upwind scheme is insignificant. However, for other mesh groups there is a significant discrepancy between solutions of the two different algorithms since numerical diffusion effects are predominant in upwind solutions. The smearing of sharp gradient is at the maximum when prism and tetrahedral mesh types are combined with first-order upwind discretization scheme. On the other hand, while hexahedral and prism type meshes produce similar results when *limitedLinear* algorithm is employed, the tetrahedral mesh solution diverges from the sharp front. Thus, these results showed that alignment of streamlines with grid boundaries is critical in terms of numerical diffusion errors. Therefore, in the present study, the *limitedLinear* scheme was used in all simulations of FVM for consistency between all mesh types.

In FVM simulations, first stationary velocity field was obtained in the T-shape passive micromixer using the designated boundary conditions and simulation parameters given in Tables 1 and 2. Computed stationary velocity domain was later employed to run two different steady-state scalar transport simulations. Initially, the steady-state AD equation was solved by setting the molecular diffusion constant to zero and this numerical solution is used to determine the effective diffusivity, which represents the amount of average numerical (or false) diffusion generated in the micromixer (i.e., $D_{Effective} \approx D_{Numerical}$). Using this solution, the mixing efficiency was calculated at the outlet of the micromixer using Equation (5). For convenience, mixing efficiency amount calculated is defined as "false mixing" since this mixing is created by numerical diffusion errors during the simulations. In a second simulation, the steady-state AD equation was solved using the physical molecular diffusion constant of the transported scalar (i.e., $D_M = 3 \times 10^{-10}$ m^2/s) and in this case, the computed effective diffusivity shows the collective effects of the numerical and molecular diffusion in the numerical solution (i.e., $D_{Effective} \approx D_{Numerical} + D_{Molecular}$). Therefore, a comparison of these two calculated effective diffusion constants may reveal which diffusion constant is predominant in the numerical solution.

As it can be seen in Figure 12a, in which horizontal axis shows the average numerical diffusion error in a logarithmic scale and vertical axis shows false mixing for each mesh type and grid size at Re = 100, hexahedral mesh produced considerably lower numerical diffusion and accordingly less false mixing in contrast to prism and tetrahedral meshes. While the order of numerical diffusion is around 10^{-13} and false mixing is 0.5% at L_1 for hexahedral mesh, these numbers sharply increased to 10^{-9} and 10% for prism and 10^{-8} and 14% for the same mesh level. Scalar transport simulations which are conducted with tetrahedral and prism mesh types produced numerical diffusion around five and four orders of magnitude higher than that of hexahedral mesh, respectively, and the magnitude of these errors manifest themselves as false mixing at the outlet. It is clear that high amount of numerical diffusion that has occurred is mainly due to non-orthogonal alignment of velocity and grid boundaries in the computational domain since other simulation parameters were all nearly constant [13]. In unstructured tetrahedral mesh type mesh flow disorientation create significant effects even at the finest mesh level in terms of false diffusion errors in the solution, but on the other side although mesh flow alignment is not maintained in prism elements, the numerical diffusion generated and false mixing obtained are noticeably lower than tetrahedral type. In prism-type mesh, a constant mesh and flow alignment is sustained through the computational domain in contrast to randomly distributed grid elements that exist in the tetrahedral type. Besides, numerical diffusion and false mixing at the outlet increases when the grid is coarser for all mesh categories. Grid coarsening effects, however show different behavior in terms of numerical diffusion and false mixing production for all mesh groups. In other words, while hexahedral mesh exhibits a minimal numerical error increase to grid size change, as it can be seen from the mild slope between mesh levels, sharply increasing slopes in the prism and tetrahedral type elements are evidence of higher rates of increase in numerical diffusion errors.

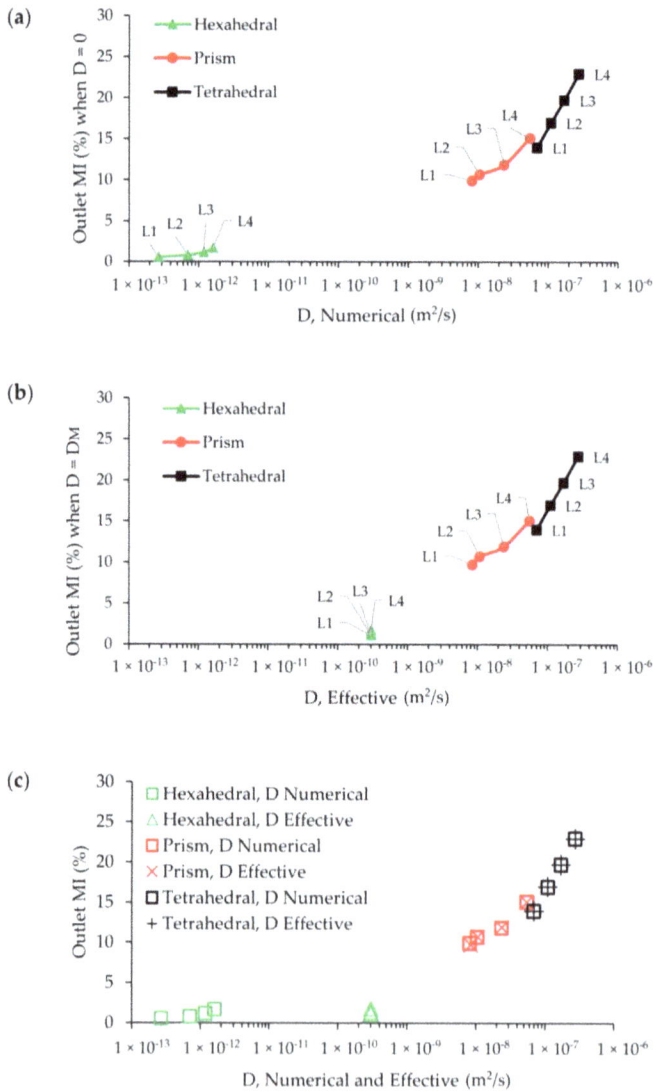

Figure 12. Numerical diffusion and effective diffusion in hexahedral, prism, and tetrahedral mesh domains at L_1, L_2, L_3, and L_4: (**a**) Numerical diffusion vs. MI; (**b**) effective diffusion vs. MI; (**c**). physical diffusion masking.

Figure 12b shows the calculated effective diffusivity and corresponding mixing efficiency at the outlet when physical diffusion is included to the solution. Calculated effective diffusivity for hexahedral mesh type is 3×10^{-10}, implying that the physical molecular diffusion constant is completely recovered from the numerical solution of the AD equation. Thus, for all levels of hexahedral mesh, the numerical solution of the AD equation reflects the effects of the physical diffusion constant as it is much higher than the numerical diffusion magnitude shown in Figure 12a. However, for other mesh types, effective diffusivity constants and false mixing values obtained at the outlet are almost equal. This is clear when Figure 12a,b is compared. If numerical diffusion and effective diffusion constants calculated from two separate simulations and the corresponding outlet mixing efficiencies are compared, as shown in Figure 12c, it is apparent that all physical diffusion effects are completely masked by numerical diffusion effects when prism and tetrahedral mesh types are used to simulate the same scalar transport scenario. This is not the case with the hexahedral mesh type. Thus, overlapped data points from two separate simulations reveal the severity of the numerical diffusion effects. Similarly, Figure 12c reveals another important point; that is the magnitude of the calculated numerical diffusions are of several orders of magnitude lower than physical molecular diffusion even for the largest grid size of the hexahedral mesh. Therefore, considering average cell Pe numbers for different grid levels of hexahedral mesh type at Re = 100 case (given in Table 2), it is possible to obtain a solute transport solution with a negligible amount of numerical diffusion even at high average cell Pe numbers. This statement may not be correct when the flow pattern in the mixing channel turns to engulfment regime since mesh flow alignment and local velocities will change and affect the numerical diffusion that may develop.

In Figure 13, the horizontal axis shows the mesh density, the right vertical axis shows numerical diffusion outcome and the left vertical axis shows the false mixing percentage obtained in the three mesh type runs. As shown in Figure 13a–c, respectively, while false mixing is less than 2% for the coarsest grid of the hexahedral mesh (e.g., around 1×10^5 cells), the false mixing is around 10% to 14% range for the prism elements and 14% to 24% range for the tetrahedral elements. We also note that for other mesh groups more than 8.5 M cells were used in the solution. When the same comparison is made for the numerical diffusion magnitudes obtained (right axis of Figure 13), one can see that while the numerical diffusion magnitude is around (10^{-12} to 10^{-13}) range for hexahedral mesh, the numerical diffusion magnitudes increases to (10^{-7} to 10^{-8}) range for prism and tetrahedral meshes. Also, curve fitting equations may be used to estimate the required mesh density to obtain a negligible amount of false mixing or numerical diffusion for the prism and tetrahedral element types. Table 4 shows the average mesh densities required to obtain predetermined false mixing and numerical diffusion values. To obtain a numerical diffusion equal to the molecular diffusion amount or to obtain 5% false mixing the required cell numbers are more than 10^9 for prism and tetrahedral mesh types. These estimates are obtained from curve fitting the data in Figure 13b,c, as shown by the dotted lines identified as (Curve (FM) and Curve (ND). These estimations reveal that the required mesh densities are beyond today's computational capacity even for the best-case scenarios in Table 4. Therefore, the results obtained confirm the previously discussed arguments that maintaining mesh flow alignment is crucial to significantly diminish numerical errors in an advection dominant transport system.

The relationship between Re number and numerical diffusion may also be investigated, as shown in Figure 14a. While the order of numerical diffusion is about 10^{-16} at Re = 0.1 for hexahedral mesh type, the magnitude of this error increases with increasing Re number and reaches to 10^{-13} for Re = 10 case. As discussed above and shown in Figure 14, the level of numerical diffusion error is several orders of magnitude less than the molecular diffusion constant of the solute for the hexahedral mesh. As the Reynolds number increases from Re = 0.1 to Re = 100, there is an increase in the numerical diffusion magnitude for the hexahedral mesh, but the even at Re = 100 the magnitude of numerical diffusion stays less than the magnitude of molecular diffusion constant of the solute. This shows that for all Re ranges considered in this study the mixing performance obtained represents the physical mixing in the T-shape mixer. As shown in Figure 14 the numerical error always increases with an increase of the Re. However, for prism and tetrahedral elements, the numerical error is at the level of the molecular diffusion level of the solute. Thus, in Figure 14b, it is shown that the combined increase of effective diffusivity which contributes the mixing performance now includes significant numerical mixing effects which would yield erroneous results in mixing percentages reported in the literature. As can be seen in Figure 14b, the effective diffusion constant for prism and tetrahedral meshes is at the level of (1×10^{-7}) which is much higher than the molecular diffusion constant of the solute. Whereas, for the hexahedral element case the increase in the effective diffusion constant is insignificant, thus all mixing performances reported would represent physical mixing conditions in the T-shape mixer.

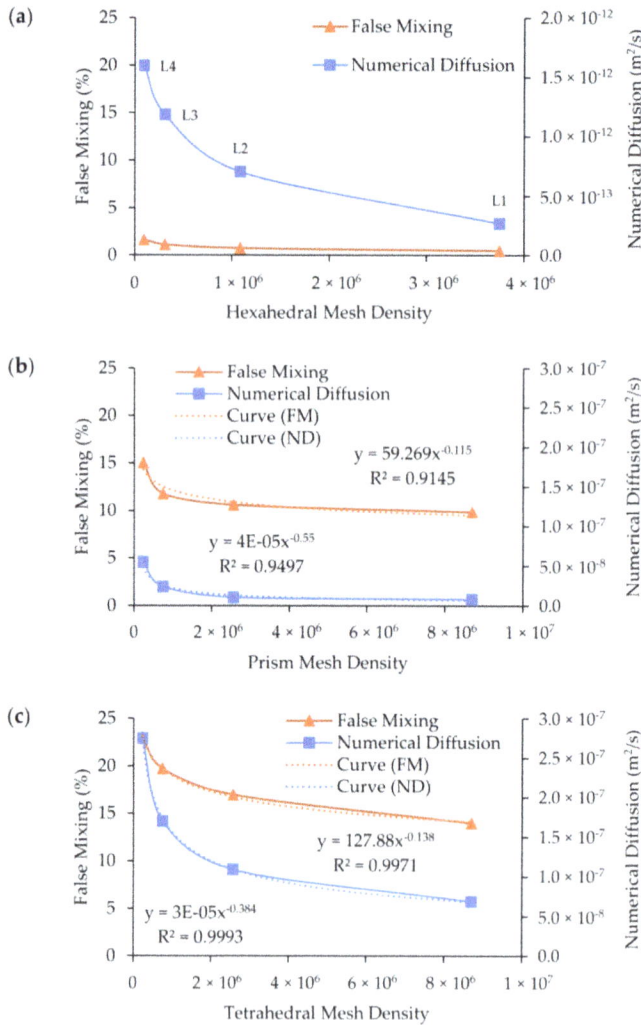

Figure 13. Change of numerical diffusion and false mixing with mesh density: (**a**) hexahedral; (**b**) prism; (**c**) tetrahedral.

Table 4. Estimation of mesh densities in the T-mixer to reach predetermined thresholds for prism and tetrahedral mesh at Re = 100 (Pe = 3.33×10^5).

False Mixing (%)	Required Average Mesh Density		Numerical Diffusion	Required Average Mesh Density	
	Prism	Tetrahedral		Prism	Tetrahedral
0.50	1.08×10^{18}	2.81×10^{17}	1.00×10^{-13}	4.37×10^{15}	1.19×10^{22}
1.00	2.61×10^{15}	1.85×10^{15}	1.00×10^{-12}	$6.64 \times 10^{13'}$	2.96×10^{19}
2.00	6.28×10^{12}	1.22×10^{13}	1.00×10^{-11}	1.01×10^{12}	7.37×10^{16}
5.00	2.18×10^{9}	1.59×10^{10}	3.00×10^{-10}	2.08×10^{9}	1.05×10^{13}

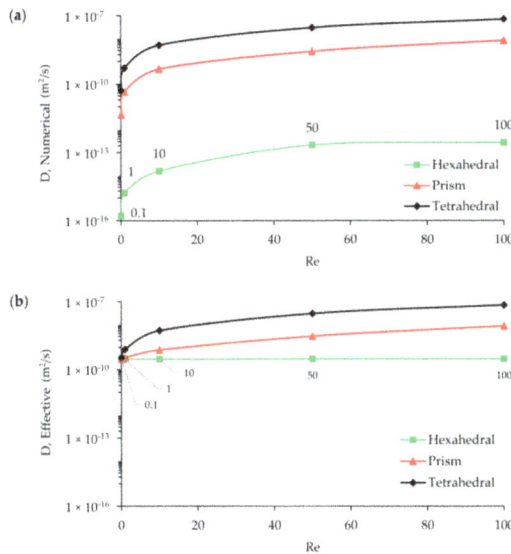

Figure 14. Numerical and effective diffusion change with Re number at L_1: (**a**) numerical diffusion vs. Re; (**b**) effective diffusion vs. Re.

These results are also summarized in Table 5 with the fractional comparisons of numerical, effective, and physical molecular diffusion constants of all mesh types and Re scenarios that are considered in this study. For all mesh types, the concentration field can be resolved with similar accuracies only at Re = 0.1 level. In addition, prism mesh provides a solution close that of hexahedral at Re = 1. However, it should be noted that although prism and tetrahedral mesh simulations provide consistent results with hexahedral type at Re = 0.1, these mesh groups use almost 2.3 times more cells than hexahedral grid type at the finest grid level. This situation will significantly affect the computational cost. For higher Re numbers, the differences in numerical errors cannot be reconciled at the mesh levels used in this study.

Table 5. Comparisons of numerical, effective, and physical diffusion constants in terms of different mesh type and Re scenarios at L_1 mesh density.

Re	D-Numerical/D-Physical			D-Effective/D-Physical		
	Hexahedral	Prism	Tetrahedral	Hexahedral	Prism	Tetrahedral
0.1	5.36×10^{-7}	0.01	0.17	1.00	1.01	1.17
1	5.39×10^{-6}	0.15	1.69	1.00	1.15	2.69
10	5.16×10^{-5}	1.53	17.03	1.00	2.53	18.03
50	6.91×10^{-4}	8.99	98.32	1.00	9.99	99.32
100	9.04×10^{-4}	26.94	230.01	1.00	27.94	231.01

Figure 15a,b show the outlet mixing efficiencies in terms of numerical and effective diffusivities for different Re scenarios at L_1 grid level. According to Figure 15a, although hexahedral mesh generates numerical diffusion with increasing Re number (or increasing average cell Pe number), these errors are several orders of magnitude less than physical diffusion constant. Thus, the transport solution does not produce a significant amount of false mixing at the outlet due to the maintained orthogonality between grids and flow. This situation is similar for Re = 0.1, 1, and 10 cases when prism and tetrahedral mesh types are employed for scalar transport solution. However, false mixing sharply increases after Re = 10 scenario since the effect of numerical diffusion increases sharply for high Re. At Re = 10, 50, and 100, the numerical diffusion produced for the prism element are on the order of 4.6×10^{-10}, 2.7×10^{-9}, and 8.0×10^{-9}, respectively. The same trend is also observed in the tetrahedral element case yielding one order of magnitude higher numerical diffusion and much higher false mixing. For these type of meshes, a greater amount of numerical diffusion generated would lead to much higher amount of false mixing at the outlet. In Figure 15b, the effective diffusion constant and outlet mixing efficiency results are shown which highlights this point. In Figure 15b, while all three mesh types provide consistent mixing efficiencies at Re = 0.1, this is only limited to smallest Re scenario because when Re number increases the behavior of three different mesh types start diverging. At Re = 0.1 effective diffusivity represents the molecular diffusion constant of transported scalar and mixing performance of the T-mixer at the outlet is around ~16% as predicted mainly for the hexahedral mesh with the other mesh types being close to this value. When the Re increases the mean residence time of the solute in T-mixer reduces which leads less mixing efficiency in the mixer since fluid particles spend less time in the mixer. As the estimated effective diffusion constants represent physical molecular diffusion in all Re scenarios for the hexahedral mesh type, for other mesh groups, effective diffusivity increases with Re number due to increasing numerical diffusion as shown in Figure 15a. Therefore, for all Re scenarios of hexahedral mesh, the real physical diffusion constant ($DM = 3 \times 10^{-10}$ m^2/s) is represented by effective diffusivity, and mixing performance decreases with increasing Re. In the case of tetrahedral mesh type, deviation from the physical molecular diffusion point ($DM = 3 \times 10^{-10}$ m^2/s) starts at Re = 1 scenario and causes around 3% more mixing at the outlet than for hexahedral mesh due to the numerical diffusion effects that are generated. In this mesh type, it is observed that although the numerical diffusion is increasing for Re = 1, 10, 50 scenarios, these effects are not reflected as physical outlet mixing since there is a compensation between residence time decrease, and thus expected decrease in mixing, and numerical mixing increase and thus mixing efficiency increase. However, this compensation balance is lost after Re = 50 since numerical diffusion effects and thus false mixing significantly increases and overcomes the effects of residence time decrease. Similar trends are also observed for tetrahedral elements but in this case the false mixing effects are much higher, as can be seen in Figure 15b.

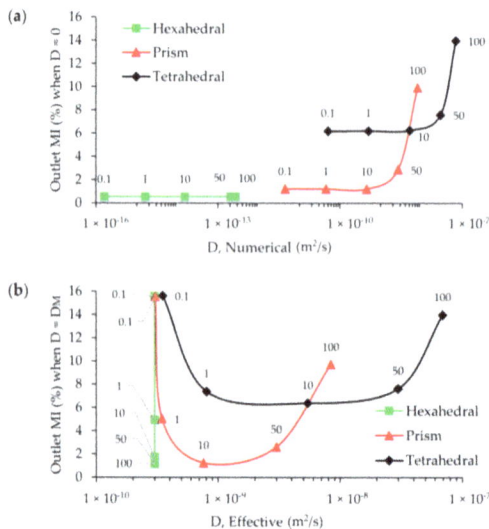

Figure 15. Mixing index change vs. numerical and effective diffusion for Re scenarios at L$_1$: (**a**) numerical diffusion vs. MI; (**b**) effective diffusion vs. MI (inserted numbers on the data points refer to Re).

4.2. FEM Analysis

In FEM analysis, simulations were only performed for L_1, L_2, L_3, and L_4 grid levels at Re = 100 and for L_1 grid level at Re = 0.1 of hexahedral mesh type using the same boundary conditions and simulation parameters with FVM. These scenarios are selected to show the effects of artificial diffusion stabilization method for advection dominant systems in FEM. When cell Pe is greater than two, the solution of scalar transport equation gives oscillatory solution in FEM depending on the magnitude of cell Pe number. In the COMSOL software, two techniques are provided to overcome instabilities during the solution of scalar transport equation that are identified as "consistent" and "inconsistent" stabilization. In consistent stabilization approach a stable solution is obtained by increasing the physical diffusion constant in the streamline direction and crosswind direction which is orthogonal to the streamlines. In the inconsistent method, the physical molecular diffusion constant is increased to reduce cell Pe number to approximately two to guarantee stability in the solution as described below.

$$\text{Pe}_\Delta = \frac{\overline{U}\Delta x}{D_M} \tag{11}$$

$$\text{Pe}_\Delta = \frac{\overline{U}\Delta x}{D_M + D_{AD}} \tag{12}$$

Equation (11) defines average cell Pe number in which $\overline{U}$ is average velocity, Δx is grid size, and D_M is molecular diffusion constant. In this study Pe_Δ is defined as average using the average velocity in the mixing channel since velocity magnitudes change at each cell in the computational domain. To decrease the average cell Pe number to a moderate number, diffusion constant is increased as described in Equation (12) in which D_{AD} is the artificial diffusion constant. It is defined as $D_{AD} = \delta\overline{U}\Delta x$, in which δ is a tuning factor, $0 < \delta < 1$. In FEM simulations, this value is selected as 0.25 and 0.50 to show artificial diffusion effects in two different magnitudes. The tuning parameters applied (0.25 and 0.50) reduce the average cell Pe number to 4 and 2, respectively, for Re = 100 scenario, and 2.22 and 1.43 for Re = 0.1 case. Similarly, the FEM also requires stabilization to resolve the flow field solution; however, as mentioned earlier, high kinematic viscosity of the fluid significantly reduces the grid Re number to moderate values at which solution does not show oscillatory trend. Figure 16a shows the velocity distribution at x = 200 μm plane in the mixing channel at Re = 100 scenario for four different mesh levels of hexahedral type. Differences between mesh levels are close to FVM solution. Also, Figure 16b confirms that both FVM and FEM resolved the flow field with an insignificant difference.

Figure 17a shows concentration distribution of the Re = 100 case at the outlet of the mixer obtained from FEM solution with consistent stabilization method. Scalar transport simulations for all different mesh levels did not show an oscillatory behavior in the solution. On the other hand, while the resolved concentration field is consistent with FVM solutions as shown in Figure 17b, a small difference between the two solutions arises at L_4 grid size. However, both methods successfully captured the sharp front in the mixing channel during the simulations. Therefore, it is obvious that consistent algorithm performs stabilization adequately.

When inconsistent stabilization is used in the simulations, results exhibit a considerable difference between consistent method as shown in Figure 18a,b. When tuning parameter is set to 0.25, which reduces the average cell Pe number to 4, sharp concentration profile significantly smears. This situation is worse when the tuning parameter is increased to 0.50 (average cell Pe = 2) and the concentration profiles are flattened noticeably. Also, the divergence in the solution increases with coarsening grid sizes. These are obviously the effects of artificially added diffusion amount. Although the solution of AD equation does not show any instabilities around cell Pe number = 2, the effects of artificial diffusion in the system are not tolerable in terms of a performance evaluation of mixing in the micromixer. As it can be examined from Equation (11), another possibility is decreasing grid size to reduce the average cell Pe number to 2; however, this method is not feasible since the required mesh density will be on the order of 10^{10}.

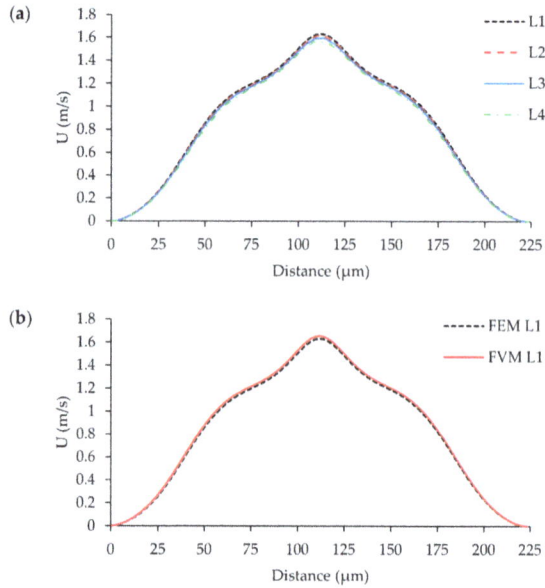

Figure 16. Velocity distribution on the plane at $x = 200$ μm for Re = 100 scenario: (**a**) different mesh levels of FEM; (**b**) comparison of FEM and FVM at L_1.

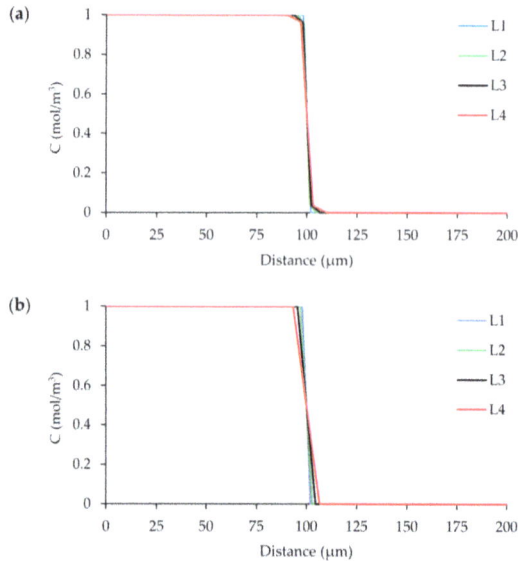

Figure 17. Outlet concentration distribution along the width of the mixing channel at $z = 50$ μm for four different grid levels at Re = 100 scenario: (**a**) FEM solution with consistent stabilization; (**b**) FVM solution.

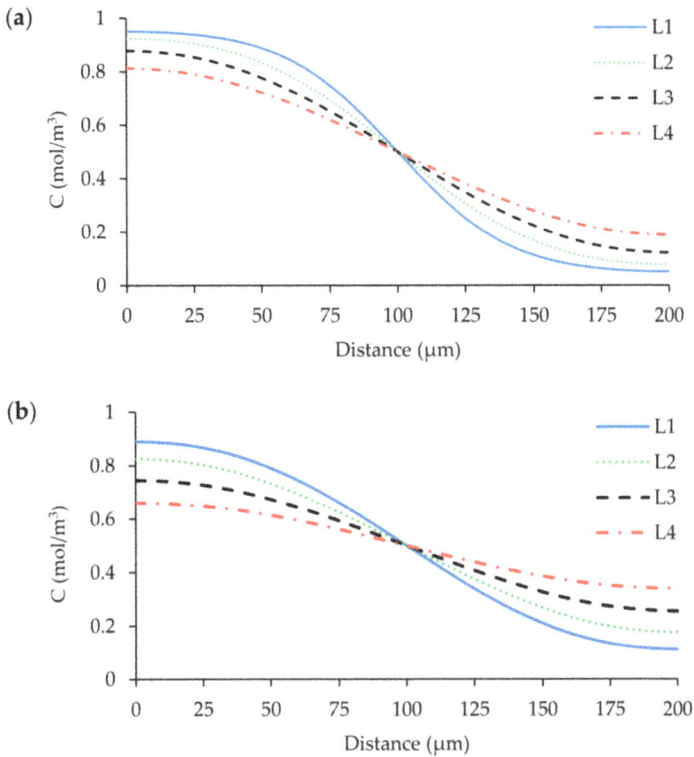

Figure 18. FEM solution outlet concentration distribution along the width of the mixing channel at $z =$ 50 µm for four different grid levels at Re = 100 scenario: (**a**) δ = 0.25; (**b**) δ = 0.50.

4.3. Comparison of FVM and FEM Solutions

When FVM and FEM are compared, a significant variance is observed between solutions in the case of an inconsistent stabilization method being used in FEM. As we plot Figures 17 and 18 together in Figure 19a, an important discrepancy emerges between the results. The inconsistent stabilization cases cannot capture the actual concentration profile at the outlet of the mixing channel and provide completely discrete concentration distribution depending on the size of the tuning factor. If the performance of the micromixer is evaluated by using the artificial diffusion method, additional diffusion in the system will completely mask the real mixing efficiency which is created by the molecular diffusion constant of the solute. As it is shown in Figure 19b, there is a negligible difference between FEM with consistent stabilization and FVM, in terms of predicted outlet mixing efficiency of the micromixer. However, when the inconsistent method is used, mixing performance of T-mixer increases up to 75% at L_4 grid size when the tuning parameter is selected as 0.50. The negative effects of the inconsistent method are still significant even for the best case (L_1 with 0.25 regulation coefficient), in which calculated mixing efficiency is more than 30%.

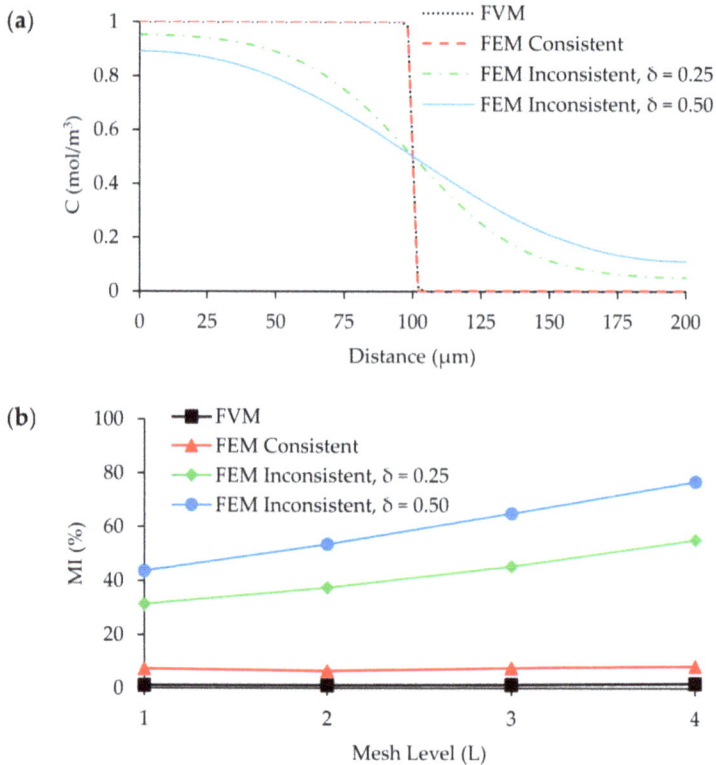

Figure 19. Comparison of FVM and FEM simulations at Re = 100: (**a**) Concentration distribution on the outlet cross section at $z = 50$ μm and L_1 grid size; (**b**) MI at the outlet for four different grid levels.

The same situation is also observed at the Re = 0.1 scenario of hexahedral mesh. FEM with consistent stabilization and FVM show almost the same concertation distribution at the outlet of the mixer and both methods provide equal amount of mixing at the outlet as shown in Figure 20a,b. On the other hand, although inconsistent methods capture different concentration profiles, the magnitude of smearing is decreased due to decreasing velocity magnitudes in the Re = 0.1 scenario. Nevertheless, the predicted outlet mixing efficiencies are still 2 (e.g., when δ = 0.25) and 2.5 (e.g., when δ = 0.50) times more than that of FVM and FEM with consistent solution. Therefore, considering all the outcomes discussed in this section, the micromixer performance may significantly change artificially depending on the FEM stabilization technique employed. By use of the consistent stabilization method, the results are almost uniform with FVM solutions; however, when the inconsistent correction method is employed in FEM, mixing efficiencies may increase depending on the magnitude of the tuning parameter selected. Thus, the stabilization method employed in FEM is crucial to reliably evaluate the mixing performance of a micromixer. In several numerical micromixer studies in which FEM is employed, the type of stabilization technique employed is not discussed and reported. However, based on the selected correction method, the mixing performance of the micromixer may be misevaluated and documented efficiencies may become susceptible relative to the physical mixing efficiency performance of the device.

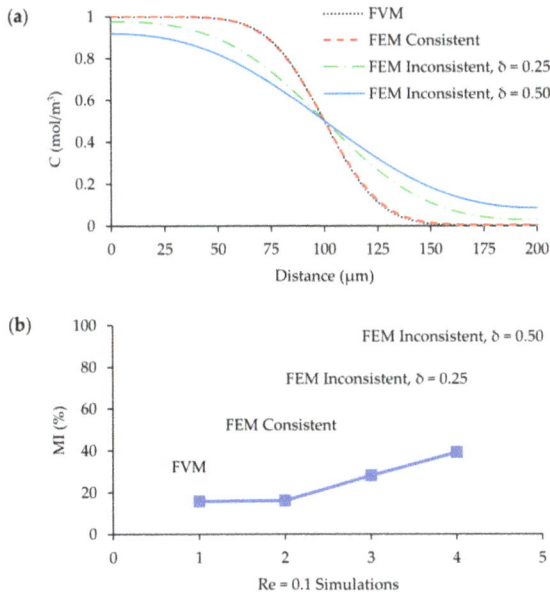

Figure 20. Comparison of FVM and FEM simulations at Re = 0.1 and L_1 grid size: (a) concentration distribution on the outlet cross section at z = 50 μm; (b) MI at the outlet.

5. Conclusions

In this study, mixing performance analysis of a T-shape micromixer was conducted using two frequently used CFD tools, namely FVM and FEM. The effects of different flow, transport, and mesh parameters were examined in terms of their effects on the evaluation of mixing performance of passive micromixers. In FVM, three types of mesh structure at four different grid levels (i.e., hexahedral, prism, and tetrahedral) were designed and different flow and transport scenarios were tested. The grid convergence analysis reported shows that mixing parameters should not be used as the criteria to evaluate grid convergence. This is because mixing parameters are significantly affected by numerical errors which may mislead the grid convergence analysis. It is shown that all mesh types selected were able to resolve the flow field accurately since all mesh groups captured the same flow profile at Re = 100 scenario. Meanwhile, keeping flow direction and grid boundaries aligned in the computational domain (hexahedral mesh case) produced a negligible amount of numerical diffusion in the solution of the AD equation for all Re cases tested in T-shape micromixer. On the other hand, disoriented mesh and flow direction caused severe numerical diffusion in the numerical solution of the concentration field at Re scenarios larger than 0.1. At the Re = 0.1 (average cell Pe number = 5) scenario, hexahedral, prism, and tetrahedral mesh types produced similar amounts of numerical diffusion and predicted almost equal amounts of physical diffusion and mixing at the outlet of the mixer. Therefore, numerical results showed that the false diffusion amount generated is mostly related to both cell Pe number and orthogonality between flow direction and grid lines. On the other hand, although the false diffusion generated and the physical mixing predicted were almost equal for all three mesh groups, prism and tetrahedral mesh types used 2.3 times more mesh elements than that of hexahedral at L_1. Thus, using the hexahedral mesh type is advantageous in terms of computational cost. This is not the case for high Re and higher grid levels in which case false diffusion effects significantly increase, masking the molecular diffusion effects. Considering the outcomes, it is recommended that prism and tetrahedral grid types should be avoided, especially for solution of the AD equation. On the other hand, FEM results showed that while the consistent stabilization method provided almost the same results with FVM for the solution of both flow and transport equations, the inconsistent method, in which the physical diffusion constant is artificially increased, severely changed the calculated concentration values and artificially increased the mixing performance of the micromixer. Therefore, when FEM is used in micromixer studies, the numerical stabilization technique employed should be reported to properly evaluate the results.

Author Contributions: Both authors contributed to the concept of the study, analysis and discussion of the results, and preparation of the paper. M.B.O. contributed to setup and completion of the simulations. M.M.A. contributed to the preparation of the paper and supervised the overall study.

Acknowledgments: This research did not receive any grants from funding agencies in the public, commercial, or not-for-profit sectors.

Conflicts of Interest: The authors declare no conflict of interest.

References

1. Mansur, E.A.; Ye, M.; Wang, Y.; Dai, Y. A state-of-the-art review of mixing in microfluidic mixers. *Chin. J. Chem. Eng.* **2008**, *16*, 503–516. [CrossRef]
2. Nguyen, N.-T.; Wu, Z. Micromixers—A review. *J. Micromech. Microeng.* **2005**, *15*, R1–R16. [CrossRef]
3. Capretto, L.; Cheng, W.; Hill, M.; Zhang, X. Micromixing within microfluidic devices. *Top. Curr. Chem.* **2011**, *304*, 27–68. [PubMed]
4. Kumar, V.; Paraschivoiu, M.; Nigam, K.D.P. Single-phase fluid flow and mixing in microchannels. *Chem. Eng. Sci.* **2011**, *66*, 1329–1373. [CrossRef]
5. Cai, G.; Xue, L.; Zhang, H.; Lin, J. A review on micromixers. *Micromachines* **2017**, *8*, 274. [CrossRef]
6. Ward, K.; Fan, Z.H. Mixing in microfluidic devices and enhancement methods. *J. Micromech. Microeng.* **2015**, *25*, 094001. [CrossRef] [PubMed]
7. Hessel, V.; Löwe, H.; Schönfeld, F. Micromixers—A review on passive and active mixing principles. *Chem. Eng. Sci.* **2005**, *60*, 2479–2501. [CrossRef]
8. Wang, L.; Liu, D.; Wang, X.; Han, X. Mixing enhancement of novel passive microfluidic mixers with cylindrical grooves. *Chem. Eng. Sci.* **2012**, *81*, 157–163. [CrossRef]
9. Alam, A.; Afzal, A.; Kim, K.-Y. Mixing performance of a planar micromixer with circular obstructions in a curved microchannel. *Chem. Eng. Res. Des.* **2014**, *92*, 423–434. [CrossRef]
10. Demirel, E.; Aral, M. Unified analysis of multi-chamber contact tanks and mixing efficiency evaluation based on vorticity field. Part II: Transport analysis. *Water* **2016**, *8*, 537. [CrossRef]
11. Rudyak, V.Y.; Minakov, A.V.; Gavrilov, A.A.; Dekterev, A.A. Modelling of flows in micromixers. *Thermophys. Aeromech.* **2010**, *17*, 565–576. [CrossRef]
12. Bailey, R.T. Managing false diffusion during second-order upwind simulations of liquid micromixing. *Int. J. Numer. Methods Fluids* **2017**, *83*, 940–959. [CrossRef]
13. Liu, M. Computational study of convective–diffusive mixing in a microchannel mixer. *Chem. Eng. Sci.* **2011**, *66*, 2211–2223. [CrossRef]
14. Moukalled, F.; Mangani, L.; Darwish, M. *The Finite Volume Method in Computational Fluid Dynamics: An Advanced Introduction with Openfoam and Matlab*; Springer Publishing Company: New York, NY, USA, 2015; p. 791.
15. Godunov, S.K.; Ryaben'kii, V.S. Spectral stability criteria for boundary-value problems for non-self-adjoint difference equations. *Russ. Math. Surv.* **1963**, *18*, 1–12. [CrossRef]
16. Idelsohn, S.R.; Oñate, E. Finite volumes and finite elements: Two 'good friends'. *Int. J. Numer. Methods Eng.* **1994**, *37*, 3323–3341. [CrossRef]
17. Patankar, S.V. *Numerical Heat Transfer and Fluid Flow*; Hemisphere Publishing Corporation: New York, NY, USA, 1980.
18. Gresho, P.M.; Lee, R.L. Don't suppress the wiggles—They're telling you something! *Comput. Fluids* **1981**, *9*, 223–253. [CrossRef]
19. Kuzmin, D. *A Guide to Numerical Methods for Transport Equations*; University Erlangen-Nuremberg: Erlangen, Germany, 2010. Available online: http://www.mathematik.uni-dortmund.de/~kuzmin/Transport.pdf (accessed on 28 April 2018).

20. Oñate, E.; Manzan, M. Stabilization techniques for finite element analysis of convection-diffusion problems. *Dev. Heat Transf.* **2000**, *7*, 71–118.
21. Leonard, B.P. A stable and accurate convective modelling procedure based on quadratic upstream interpolation. *Comput. Methods Appl. Mech. Eng.* **1979**, *19*, 59–98. [CrossRef]
22. Van Leer, B. Towards the ultimate conservative difference scheme. V. A second-order sequel to Godunov's method. *J. Comput. Phys.* **1979**, *32*, 101–136. [CrossRef]
23. Soleymani, A.; Kolehmainen, E.; Turunen, I. Numerical and experimental investigations of liquid mixing in T-type micromixers. *Chem. Eng. J.* **2008**, *135*, S219–S228. [CrossRef]
24. Le Thanh, H.; Dong, T.; Ta, B.Q.; Tran-Minh, N.; Karlsen, F. An effective passive micromixer with shifted trapezoidal blades using wide reynolds number range. *Chem. Eng. Res. Des.* **2015**, *93*, 1–11. [CrossRef]
25. Virk, M.S.; Holdø, A.E. Numerical analysis of fluid mixing in T-type micro mixer. *Int. J. Multiphys.* **2016**, *2*. [CrossRef]
26. Jain, M.; Rao, A.; Nandakumar, K. Numerical study on shape optimization of groove micromixers. *Microfluid. Nanofluid.* **2013**, *15*, 689–699. [CrossRef]
27. Gidde, R.R.; Pawar, P.M.; Ronge, B.P.; Misal, N.D.; Kapurkar, R.B.; Parkhe, A.K. Evaluation of the mixing performance in a planar passive micromixer with circular and square mixing chambers. In *Microsystem Technologies*; Springer Publishing Company: New York, NY, USA, 2017.
28. Li, T.; Chen, X. Numerical investigation of 3D novel chaotic micromixers with obstacles. *Int. J. Heat Mass Transf.* **2017**, *115*, 278–282. [CrossRef]
29. Galletti, C.; Roudgar, M.; Brunazzi, E.; Mauri, R. Effect of inlet conditions on the engulfment pattern in a t-shaped micro-mixer. *Chem. Eng. J.* **2012**, *185–186*, 300–313. [CrossRef]
30. Roudgar, M.; Brunazzi, E.; Galletti, C.; Mauri, R. Numerical Study of Split t-Micromixers. *Chem. Eng. Technol.* **2012**, *35*, 1291–1299. [CrossRef]
31. Bothe, D.; Stemich, C.; Warnecke, H.-J. Fluid mixing in a t-shaped micro-mixer. *Chem. Eng. Sci.* **2006**, *61*, 2950–2958. [CrossRef]
32. Tran-Minh, N.; Dong, T.; Karlsen, F. An efficient passive planar micromixer with ellipse-like micropillars for continuous mixing of human blood. *Comput. Methods Programs Biomed.* **2014**, *117*, 20–29. [CrossRef] [PubMed]
33. Tseng, L.-Y.; Yang, A.-S.; Lee, C.-Y.; Hsieh, C.-Y. Cfd-based optimization of a diamond-obstacles inserted micromixer with boundary protrusions. *Eng. Appl. Comput. Fluid Mech.* **2011**, *5*, 210–222. [CrossRef]
34. Van Doormaal, J.P.; Raithby, G.D. Enhancements of the simple method for predicting incompressible fluid flows. *Numer. Heat Transf.* **1984**, *7*, 147–163. [CrossRef]
35. Warming, R.F.; Beam, R.M. Upwind second-order difference schemes and applications in aerodynamic flows. *AIAA J.* **1976**, *14*, 1241–1249.
36. Sweby, P.K. High resolution schemes using flux limiters for hyperbolic conservation laws. *SIAM J. Numer. Anal.* **1984**, *21*, 995–1011. [CrossRef]
37. Geuzaine, C.; Remacle, J.F. Gmsh: A 3-D finite element mesh generator with built-in pre- and post-processing facilities. *Int. J. Numer. Methods Eng.* **2009**, *79*, 1309–1331. [CrossRef]
38. Freitas, C.J. The issue of numerical uncertainty. *Appl. Math. Model.* **2002**, *26*, 237–248. [CrossRef]
39. Celik, I.B.; Ghia, U.; Roache, P.J. Procedure for estimation and reporting of uncertainty due to discretization in CFD applications. *J. Fluids Eng.* **2008**, *130*, 078001.
40. Roache, P.J. Quantification of uncertainty in computational fluid dynamics. *Annu. Rev. Fluid Mech.* **1997**, *29*, 123–160. [CrossRef]
41. Roache, P.J. Perspective: A method for uniform reporting of grid refinement studies. *J. Fluids Eng.* **1994**, *116*, 405–413. [CrossRef]
42. Afzal, A.; Kim, K.-Y. Passive split and recombination micromixer with convergent–divergent walls. *Chem. Eng. J.* **2012**, *203*, 182–192. [CrossRef]
43. Javaid, M.; Cheema, T.; Park, C. Analysis of passive mixing in a serpentine microchannel with sinusoidal side walls. *Micromachines* **2018**, *9*, 8. [CrossRef]

micromachines

MDPI

Article

Optimization of Wavy-Channel Micromixer Geometry Using Taguchi Method †

Nita Solehati [1], Joonsoo Bae [1],* and Agus P. Sasmito [2]

[1] Department of Industrial and Information Systems Engineering, Chonbuk National University,
    567 Baekje-daero, Deockjin-gu, Jeonju 54896, Korea; nita.solehati@gmail.com
[2] Department of Mining and Materials Engineering, McGill University, Adams Building #115,
    3450 University Street, Montreal, QC H3A 2A7, Canada; ap.sasmito@gmail.com or agus.sasmito@mcgill.ca
*   Correspondence: jsbae@jbnu.ac.kr; Tel.: +82-10-2019-8306
†   This paper is an extended version of our paper published in the FAIM 2013–23rd International Conference
    on Flexible Automation & Intelligent Manufacturing, was held in Porto, Portugal, 26–28 June 2013.

Received: 17 January 2018; Accepted: 2 February 2018; Published: 6 February 2018

Abstract: The micro-mixer has been widely used in mixing processes for chemical and pharmaceutical industries. We introduced an improved and easy to manufacture micro-mixer design utilizing the wavy structure micro-channel T-junction which can be easily manufactured using a simple stamping method. Here, we aim to optimize the geometrical parameters, i.e., wavy frequency, wavy amplitude, and width and height of the micro channel by utilizing the robust Taguchi statistical method with regards to the mixing performance (mixing index), pumping power and figure of merit (FoM). The interaction of each design parameter is evaluated. The results indicate that high mixing performance is not always associated with high FoM due to higher pumping power. Higher wavy frequency and amplitude is required for good mixing performance; however, this is not the case for pumping power due to an increase in Darcy friction loss. Finally, the advantages and limitations of the designs and objective functions are discussed in the light of present numerical results.

Keywords: geometrical design; micromixer; optimum; Taguchi; wavy-channel

1. Introduction

Recent advances in micro-reactor technologies have enabled chemical processes and pharmaceutical industries to produce high quality products due to their ability to control the extreme/unusual reaction environments, such as highly exothermic or explosive chemical reaction, highly viscous fluids which are difficult to mix in larger scale mixing equipment, etc. There are many other advantages of micro-reactor technology, such as higher transport rate, safer environment, compact design and simpler process control. Despite its advantages, micro-reactor also has limitations, especially when large throughput product in industrial scale is desired, whilst a small size micro-reactor can only produce a small amount of yield, and enlarging the micro-reactor size (scaling-up) decreases the product quality. One way to increase the product output is by numbering-up the micro-reactors into several modulars; the modular comprises several mixing zones (mixers) and reaction zones (reactors). However, one of the major drawbacks of this design is the flow uniformity, for which the reactant flow may not be uniformly distributed throughout each micro-channel, which causes non-uniform product quality. Liu et al. [1] proposed the structural bifurcation of the flow channel to improve flow uniformity throughout the micro-reactor/micro-mixer modular.

The conventional micro-mixer design typically uses a T-junction with straight micro-channel: the T-junction consists of at least two inlets for the reactant to enter and further mix in the straight micro-channel. This design is widely used in chemical and pharmaceutical industries due to its ease of manufacture. However, this design has poor mixing quality, as small-scale mixing depends mainly on the molecular diffusion. Thus, a relatively long channel and higher pumping power is required to achieve desired mixing, which can be impractical. Several designs have been proposed by many researchers (see [2–6] for reviews of these) to improve the mixing quality. However, most of the proposed passive mixer designs require complex geometrical structures which are difficult/expensive to manufacture. Active mixers, on the other hand, have been proven to significantly enhance mixing and reaction rate; for example, the use of acoustic waves in sidewall sharp-edges [7,8], acoustic waves on Y-cut 128° LiNbO₃ [9] and the use of lateral acoustic transducer in micromixer [10]. Despite its advantages, the active mixer requires additional equipment, such as a wave generator, external piezoelectric buzzer and so forth, which adds capital and operating costs as well as complexity. For the industrial scale of cheap and mass-produced chemical product, active mixers increase the total production cost and may not be preferred. Hence, a simple, cheap, reliable and high performance passive micromixer is desired by such industry.

Currently, there are many available micromachining methodologies for rapid microchannel manufacturing; these include the casting of laser ablation, hot embossing, polydimethylsiloxane (PDMS), micropowder blasting, stereolithography, micromilling, and lamination. Lamination of polymeric films is another fast prototyping process for thermoplastic-based microfluidic devices. However, as it is mostly performed by laser cutting and lamination of thin layers of plastic, it requires expensive equipment and has a high maintenance cost. While the most advanced micomanufacturing technologies enable the creation of complicated three-dimensional microchannel structures, they are, however, only feasible to be implemented for high value product in their current stage due to their complication processes and expensive cost. Tonkovich et al. [11] showed that the most economical method for mass manufacturing of microchannel reactors is the stamping method; however, this method is unable to create complex geometry in three-dimensional shapes as proposed by many researchers to enhance mixing performance. Recently, the stamping method has been successfully used to fabricate cooling channels [12], microfluidic on sheath flow [13], and gas channels in proton exchange membrane fuel cells [14]. Choi et al. [15] showed that PDMS stamp can be used to fabricate micro- and nanopatterns. Thus far, the stamping method has been proven to be very economical for this purpose. The only foreseen possible drawback is the precise control of the stamping force that might not be uniform throughout the channel, and thus may create a slight error in providing the desired uniform channel height.

Numerous experimental and numerical studies to evaluate mixing enhancement have been performed. Hossain et al. [16] proposed two-layer serpentine crossing channels for mixing enhancement at low Reynolds numbers. In another study, Hossain and Kim [17] introduced the concept of the three-dimensional serpentine split-and-recombine (SAR) microchannel using a series of "OH"-shaped segments which showed a mixing index of 0.884 at Re = 30. Ahmed and Kim [18] evaluated the effect of geometrical parameters on an electro-osmotic micromixer with heterogeneous charged surface patches. Xie and Xu [19] simulated an oscillating feedback micromixer comprising an inlet channel, two Coanda steps, a divergent chamber, a splitter, two feedback channels, and an outlet channel was designed considering the Coanda effect. The results indicated that the mixing efficiency increased with the increase of Reynolds number, and a mixing efficiency of 75.3% could be achieved at Re = 100. Despite the wide range of studies performed to come up with the most efficient micromixer geometry, most of the proposed geometries have a difficult shape to be mass-manufactured.

In previous work [20], we introduced a micro-mixer with wavy structure for improved mixing performance but which was easy to manufacture by using the stamping method, similar to conventional micro-mixer design. We also investigated the details of the flow and mixing behaviour in this type of new micro-mixer design [21]; it was found that our new design of micro-mixer with wavy structure has superior performance as compared to conventional design and has comparable performance with micro-mixers with complex geometrical structures which are difficult and/or expensive to manufacture. However, for the best and better performance of our micro-mixer design with wavy structure, optimization of geometrical design is required, which is the theme of this work.

In previous work, we had successfully utilized the computational fluid dynamics (CFD) approach together with the Taguchi statistical method to optimize the performance of liquid-cooled fuel cells [22] and open-cathode fuel cells [23]. To continue our work in the area of numerical micro-mixer design and optimization using Taguchi statistical method, the aim of the study presented is twofold: (i) to optimize geometrical design—wavy frequency, wavy amplitude, micro-channel width and height; (ii) to evaluate the objective function of optimization with regards to the mixing performance (mixing index), parasitic load (pumping power) and figure of merit for different industrial application.

The layout of the paper is as follows. First, the model development using computational fluid dynamics (CFD) is introduced; it comprises conservation equations of mass, momentum and species for mixing. The mathematical model is then solved numerically utilizing finite-volume-based CFD software ANSYS Fluent 16 (Ansys, Inc., Canonsburg, PA, USA). The mixing performance of the conventional T-junction design is compared with wavy channel and complicated channel design along with parametric study on geometrical parameters. The Taguchi statistical method is then employed to study the sensitivity of each design parameter. Optimum parameters are then calculated based on the mixing performance, pumping power and figure of merit defined later. Finally, advantages and limitations of the design are highlighted, and conclusions are drawn based on the results presented.

2. Model Development

The physical model (see Figure 1) comprises a micro-wavy-channel design for which liquid A enters the channel from the right inlet (red arrow in Figure 1), while liquid B flows from the left inlet (blue arrow in Figure 1). Liquids A and B mix in the opposing streams in a T-junction. The channel height (h), width (w), wavy frequency (f) and wavy amplitude (a) are varied according to the Taguchi array. For comparison purposes, we keep the length the same for all cases.

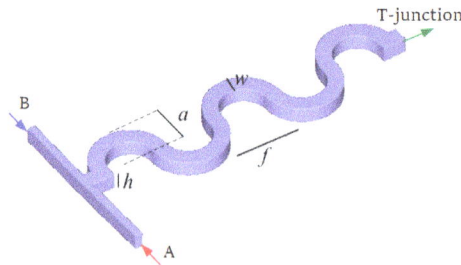

Figure 1. Schematic of micromixer T-junction with wavy structure.

2.1. Governing Equations

The conservation equations of mass, momentum and miscible species are given by

$$\nabla \cdot \rho \boldsymbol{u} = 0 \tag{1}$$

$$\nabla \cdot \rho \boldsymbol{u} \times \boldsymbol{u} = -\nabla p + \nabla \cdot [\{\mu(\nabla \boldsymbol{u} + (\nabla \boldsymbol{u})^{T})\}] \tag{2}$$

$$\nabla \cdot (\rho \boldsymbol{u} \omega_{i}) = -\nabla \cdot (\rho D_{i} \nabla \omega_{i}) \tag{3}$$

In the above equations, ρ is the fluid density, $\boldsymbol{u}$ is the fluid velocity, p is the pressure, μ is the dynamic viscosity, ω_{i} is the mass fraction of species i, D_{i} is the diffusion coefficient of species i.

The mixing performance is evaluated using mixing index, defined as

$$\tau^{2} = \frac{1}{n} \sum_{i=1}^{n} (\omega_{i} - \omega_{\infty})^{2} \tag{4}$$

$$M_i = 1 - \sqrt{\frac{\tau^2}{\tau_{max}}} \tag{5}$$

where τ indicates the variation of concentration for each cross section, τ_{max} is the maximum variance over the range of data, n is the number of sampling points inside the cross-section, ω_i is the mass fraction at sampling point i, ω_0 is the initial concentration, ω_∞ is the concentration at infinity, and M_i is the mixing index. The mixing index is unity for complete mixing, and zero for no mixing. The values at the sampling points were obtained by interpolation with the values from adjacent computational cells.

To ensure the fidelity of comparison for both micromixer designs, the concept for figure of merit (FoM) is introduced to evaluate the effect of Reynolds number and the effect of geometry on the pressure drop and mixing performance. FoM is defined as the ratio of the mixing index per unit pressure drop required, given by:

$$\text{FoM} = \frac{M_i}{\Delta p} \tag{6}$$

2.2. Boundary Conditions

The boundary conditions for the flow inside the micro-channel T-junction are as follows:

Right inlet: liquid A is introduced to the channel; we prescribe inlet velocity and species mass fraction.

$$u = U_A, \ \omega_A = 1, \omega_B = 0 \tag{7}$$

Left inlet: liquid B enters the channel; constant inlet velocity and species mass fraction are prescribed.

$$u = U_B, \ \omega_A = 0, \omega_B = 1 \tag{8}$$

Outlet: we specify the pressure and stream-wise gradient of the temperature, and species mass fraction is set to zero. The velocity is not known a priori but needs to be iterated from the neighboring computational cells.

$$p = p_{out}, \ \boldsymbol{n} \cdot \nabla \omega_i = 0 \tag{9}$$

At the walls: we specify no slip condition and no species flux at the channel wall.

$$u = 0, \ \nabla \omega_i = \boldsymbol{0} \tag{10}$$

2.3. Taguchi Statistical Method

The Taguchi method is a well-known statistical method developed by Genichi Taguchi. It is a powerful engineering tool for experimental optimization and one of the most well-known robust design methods. Generally, it is used to find the sensitivity of each parameter and determine the optimum combination of the design factors [24,25]. Here, we have four key parameters, e.g., wavy frequency, wavy amplitude, channel width and channel height, with three level values for each parameter. An L9 orthogonal array (OA) was employed in the experiment matrix, as shown in Table 1. It is worth mentioning that if one would like to investigate the effect of combination of parameters and optimize the design without the Taguchi statistical method, the total number of simulations would have been prohibitive; for example, as we have four parameters and three levels, the total number of simulation is $4^3 = 64$ simulations. In a wavy micromixer, the mesh size needs to be very fine and computationally expensive. The statistical method, on the other hand, can be used to effectively reduce the number of simulations and computational cost, evaluate the interaction of parameters and optimize the design.

In this paper, we evaluate the objective function of the optimum parameters based on the mixing index, Equation (5), pumping power (pressure drop) and figure of merit, Equation (6); therefore, we evaluate the signal-to-noise (S/N) ratio based on the-larger-the-better for mixing index and FoM:

$$S/N = -10 log_{10} \left(\frac{1}{n_r} \sum_{i=1}^{n_r} \frac{1}{Y_i^2} \right) \tag{11}$$

Whereas, for pumping power, we calculate the signal-to-noise (S/N) ratio based on the-smaller-the-better:

$$S/N = -10 log_{10} \left(\frac{1}{n_r} \sum_{i=1}^{n_r} Y_i^2 \right) \tag{12}$$

Once the optimum combination of each parameter has been determined, we verify the predicted results from Taguchi method with CFD results. The confidence interval (CI) of the estimated value is calculated by:

$$CI = \sqrt{F_{\alpha,v_1,v_2} V_{ep} \left(\frac{1}{n_{eff}} + \frac{1}{r} \right)} \tag{11}$$

where F_{α,v_1,v_2} is the F-ratio required, v_1 is the number of degree of freedom of the mean, v_2 is the number of degree freedom of the error, V_{ep} is the error of variance, r is the sample size in the confirmation test, and n_{eff} is the effective sample size, defined as

$$n_{eff} = \frac{N}{1 + DOF_{opt}} \tag{12}$$

where N is total number of trials and DOF_{opt} is the total degree of freedom that are associated with items used to estimate η_{opt}.

Table 1. Orthogonal array for L$_9$ with four parameters and three level designs.

No.	Frequency (π)	Amplitude (mm)	Width (mm)	Height (mm)
1	2	0.5	0.25	0.25
2	2	1	0.5	0.5
3	2	2	1	1
4	5	0.5	0.5	1
5	5	1	1	0.25
6	5	2	0.25	0.5
7	10	0.5	1	0.5
8	10	1	0.25	1
9	10	2	0.5	0.25

3. Numerics

The computational domains (see Figure 1) were created in AutoCAD 2010 (Autodesk, Inc., San Rafael, CA, USA); the commercial pre-processor software Gambit 2.3.16 (Ansys, Inc., Canonsburg, PA, USA) was used for meshing, labeling boundary conditions and to determine the computational domain. Three different mesh designs—1×10^7, 2×10^7 and 4×10^7—were implemented and compared in terms of the local pressure, velocities, species mass fractions and temperatures to ensure a mesh independent solution. We found that the mesh numbers around 2×10^7 give about 1% deviation compared to a much finer mesh size of 4×10^7; whereas the results from the mesh size of 1×10^7 deviate up to 10% as compared to those from the finest mesh design. Therefore, a mesh consisting of around 2×10^7 elements was found to be sufficient for the numerical experiments: a fine structured mesh was used near the wall to resolve the boundary layer and an increasingly coarser mesh in the middle of the channel in order to reduce the computational cost. The aspect ratio of the boundary layer mesh is 2, with a thickness of about 10 to 20% of the total height/width. The detail validation of the mesh independence test was published in the earlier paper [20]; for the sake of brevity, we do not repeat in this paper.

The equations were solved with the well-known semi-implicit pressure-linked equation (SIMPLE) algorithm, first-order upwind discretization and algebraic multi-grid (AMG) method. As an indication of the computational cost, it is noted that, on average, around 5000–10,000 iterations are needed for convergence criteria for all relative residuals of 10^{-9}; this takes around two days on a computer cluster with 16 processors and 20 GB of random access memory (RAM).

The key operating parameters are then analyzed using the Taguchi statistical method in Minitab 14 software. A variance analysis (ANOVA) was performed in order to see the sensitivity of each parameter, to determine the optimum combination of operating parameters and to evaluate the confidence levels between the Taguchi prediction and CFD results.

4. Results and Discussion

The numerical simulations were carried out for typical conditions found in micro-channel T-junctions; the base-case conditions together with the physical parameters and geometric parameters are listed in Table 2 which were chosen based on the typical microchannel geometries. Model verification was carried out in earlier publication [20] and for the sake of brevity is not repeated here. In the following, sensitivity analysis of each design parameter is investigated from the response of signal-to-noise ratio of OA. Optimum design parameters are then examined based on the mixing index, pressure drop (pumping power) and figure of merit (FoM).

Table 2. Physical and geometrical parameters.

Parameter	Value	Unit
Channel length	10	mm
Liquid density	998	kg/m^3
Viscosity	1×10^{-3}	kg/m·s
Diffusivity	2.2×10^{-9}	m^2/s
Velocity inlet A	0.04	m/s
Velocity inlet B	0.04	m/s

4.1. Effect of Channel Geometries

One of the key factors that determine the mixing and reaction performance is the geometric design of the channel. This study examines three different micro-channel T-junction geometries: straight, complex 3D serpentine channel and wavy channel with same total length and fluid velocity. Since the mixing is directly linked to the flow behavior and total mixing time, it is of interest to investigate the flow patterns inside the channel. Previous work on wavy microchannels [20] showed that the presence of centrifugal force due to curvature leads to significant radial pressure gradients in the flow core region. In the proximity of the inner and outer walls of the coils, however, the axial velocity and the centrifugal force approach to zero. Hence, to balance the momentum transport, secondary flow should develop along the outer wall. This is indeed the case, as can be seen in Figure 2, where the secondary flows present in the complex 3D serpentine channel and wavy channel (Figure 2b,c). This, however, is not the case for the straight T-junction, as a fully developed flow exists inside the channel. It is noted that, at this particular inlet Reynolds number (~10, Re = $\rho U Dh/\mu$), the secondary flows appear as two pairs in wavy channel as the wavy structure inverted the secondary flow direction.

The presence of secondary flow with higher velocities toward the outer wall of the complex 3D serpentine and wavy channels is expected to have direct impact on mixing characteristics. This can be inferred from Figure 2, which presents local mass fraction distribution of liquid mixing over the cross sections of various channel designs. Here, several features are apparent; foremost is that, for the same total mixing time, the complex 3D serpentine channel yields the best mixing performance with mixing index of nearly perfect 1 (0.999). The conventional straight channel T-junction, on the other hand, performs the worst among the others, with a mixing index of 0.071. On closer inspection, we note that the mixing is very poor for the straight channel T-junction as the liquid is hardly mixing throughout the channel outlet, as can be seen in Figure 2a, while the mixing performance of the wavy microchannel lies between the conventional straight T-junction and complex 3D serpentine channel with mixing index of 0.476. This indicates that the wavy microchannel has the potential to be used as a passive mixer, as the manufacturing cost and processes are much simpler than that of the complex 3D serpentine design; of course, further geometry optimization is required for best performance.

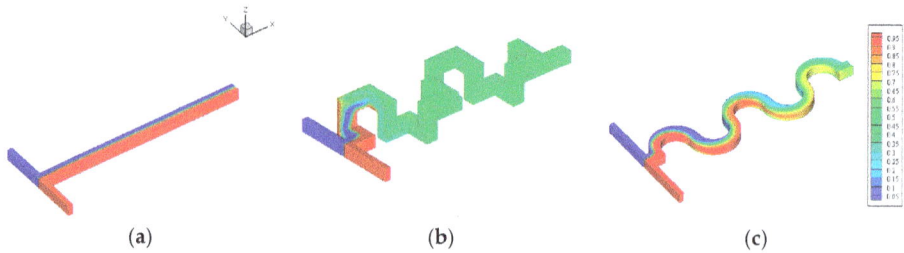

(a)   (b)   (c)

Figure 2. Mixing performance of micromixer T-junction designs. (**a**) Conventional straight; (**b**) Complex 3D serpentine; (**c**) Wavy microchannel.

4.2. Mixing Performance

This study examines the mixing performance based on OA of the Taguchi method, which is tabulated in Table 3 and for which the best mixing is achieved by design number 9 and the worst performance is given by design number 2. The sensitivity of each parameter is then analyzed by employing analysis of variance (ANOVA). Typically, for high quality and expensive chemical and/or pharmaceutical products, product quality—in this regard reflected by mixing quality—is of paramount importance. Thus, the objective function of our optimization is based on the mixing index. The higher the mixing index, the better the mixing quality.

Table 3. Numerical results of various combination of design factors.

No.	Mixing Index	Pressure Drop (Pa)	Figure of Merit
1	0.171	286.76	5.98×10^{-4}
2	0.105	87.2	1.21×10^{-3}
3	0.149	41.85	3.56×10^{-3}
4	0.115	68.36	1.68×10^{-3}
5	0.142	238.73	5.93×10^{-4}
6	0.563	3747.13	1.50×10^{-4}
7	0.203	103.58	1.96×10^{-3}
8	0.582	3506.95	1.66×10^{-4}
9	0.649	13,729	4.73×10^{-5}

Earlier work [12] showed that wavy amplitude and frequency play a significant role in the mixing performance; this is indeed the case, as can be inferred from Figure 3, where wavy frequency results in the most significant parameter influencing the mixing performance, followed by wavy amplitude. Higher wavy frequency and longer wavy amplitude improve mixing performance. This can be adequately explained by the fact that, at higher wavy frequency and longer wavy amplitude, the secondary flow generated by curve geometry is stronger, which enhances mixing. Further, increasing frequency and amplitude increases the total microchannel length, which in turn increases the residence time of the fluid mixing. On the other contrary, the width of the microchannel has a less significant effect, while microchannel height has the least significant effect on the mixing performance. We note that mixing performance improves as the width and height are decreased. This is attributed to the smaller channel dimension which reduces the diffusion path between two fluids to penetrate each other and mix.

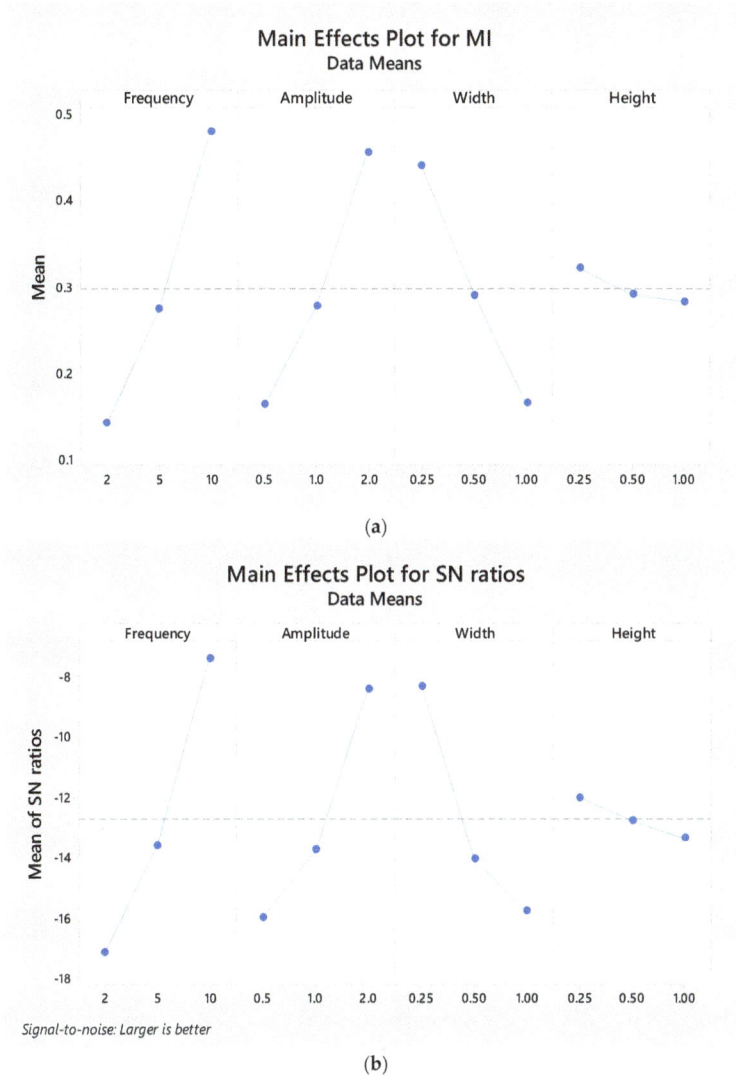

(a)

(b)

Figure 3. Taguchi results of (**a**) Mean and (**b**) signal-to-noise (S/N) response graph for mixing performance.

Figure 4 shows an interaction plot for each factor, for which parallel plot denotes no interaction while crossing indicates significant interaction. Interestingly, although channel height yields the least significant individual effect to the mixing performance, it shows the strongest interaction with other parameters, as shown by the crossing lines in Figure 4. It is worth mentioning that channel height determines the characteristic length of the channel which defines the Reynolds number and, thus, the mixing flow regimes, i.e., segregated, vortex, engulfment and chaotic flow, and mixing performance [12].

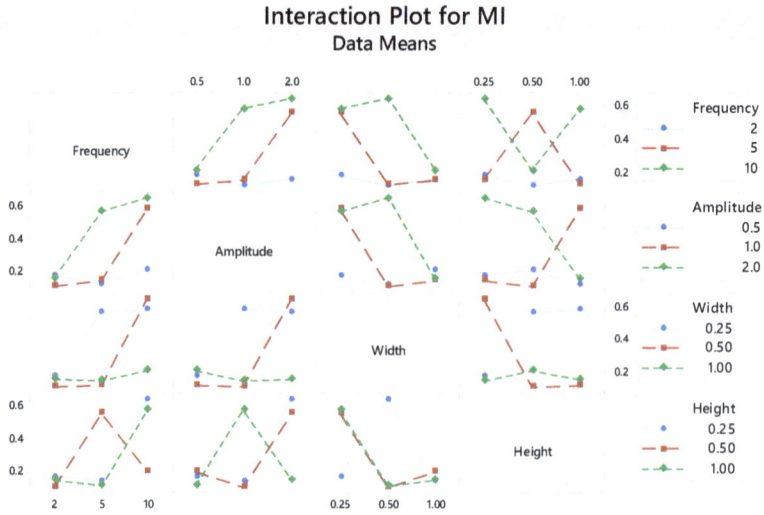

Figure 4. The interactions of various parameters with respect to mixing index.

Thus far, the sensitivity of each parameter has been examined. Now, the optimum combination of design parameters is determined. We further predict the optimum mixing performance using the Taguchi method and run the confirmatory test from CFD model. The results are depicted in Table 4 for which good agreement between the Taguchi prediction and CFD mode was obtained within the maximum error of less than 6%, which is sufficient for engineering purposes. The optimum mixing index for the optimized design is found to be 0.8.

Table 4. Optimum combination of design factors.

Parameter	Mixing Index	Pressure Drop	Figure of Merit
Frequency	10	2	2
Amplitude	2	0.5	0.5
Width	0.25	1	1
Height	0.25	1	1
Optimized design	0.8	21.75	3.77×10^{-3}
CI (%)	94.6%	93.8	95.8

4.3. Pumping Power

Generally, in cheap and mass production of chemical products, production cost becomes the most significant factor. One of the factors that constitute the production cost is the pumping power required to drive the flow mixing. Here, we evaluate the pumping power by looking at the pressure drop required. In essence, to reduce production cost, the pumping power which is represented by pressure drop should be as low as possible. The results for pressure drop required of OA are summarized in Table 3, column 3. It is seen that the highest pressure drop required is in design number 9, which is about three order-of-magnitudes higher than that of design number 3 which yields the lowest pressure drop. It is important to note that design number 9 has higher mixing performance with the expense of a much higher pressure drop. Thus, for cheap and mass production chemical product, this design seems to be not attractive, as they would prefer to implement the design with the lowest pumping power to save power/electricity cost.

With regard to the pumping power, the sensitivity of each design parameter is evaluated. Figure 5 shows the behavior of each parameter, which is somewhat different than when it was evaluated in term of mixing performance. Here, several features are apparent; foremost among them is that the low pressure drop can be obtained at low wavy frequency, low wavy amplitude, longer width and longer height, which is opposite to mixing performance. This is due to the fact that reducing frequency and amplitude reduces flow resistance throughout the microchannel as the tube length has a proportional relation with pressure drop due to the increase in Darcy friction loss ($\Delta p = f_D \times L/D \times \rho U^2/2$).

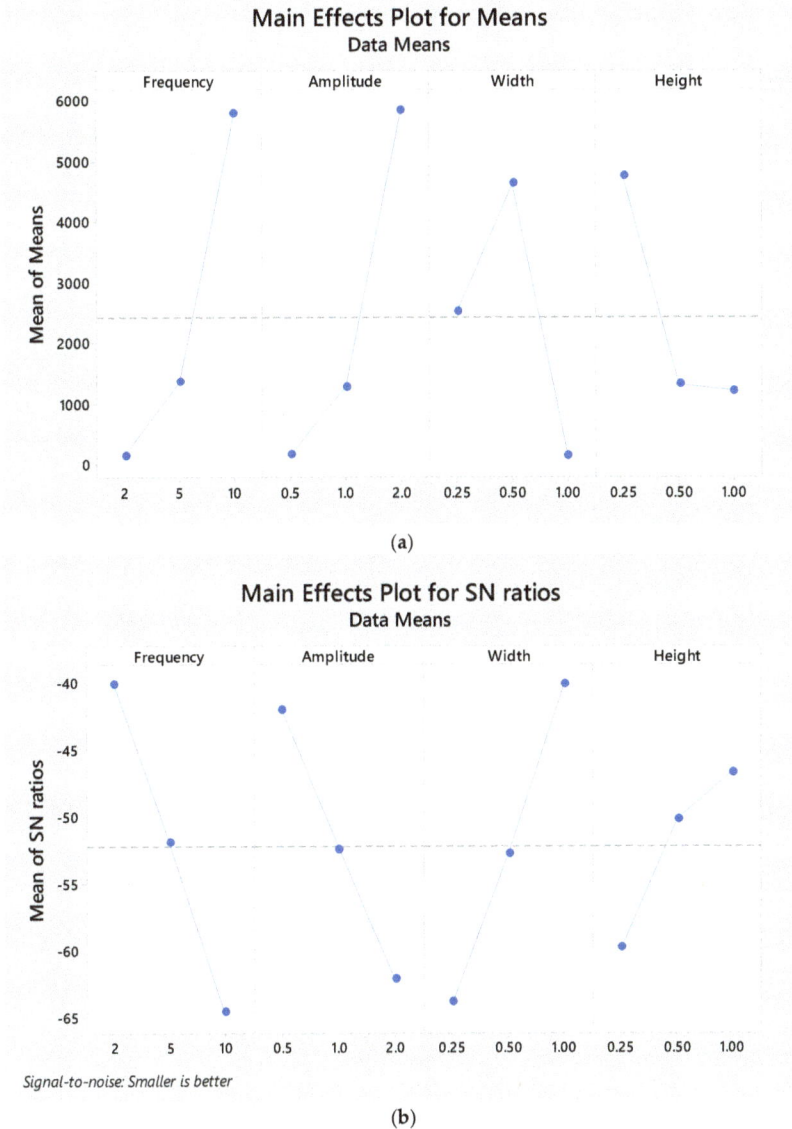

(a)

Signal-to-noise: Smaller is better

(b)

Figure 5. Taguchi results of (**a**) Mean and (**b**) S/N response graph for pumping power.

Looking at the interaction of each factor in Figure 6, it is found that significant interaction is obtained between channel height and frequency, channel width and amplitude, and channel height and width, which is reflected by interaction of L and D in Darcy friction loss equation.

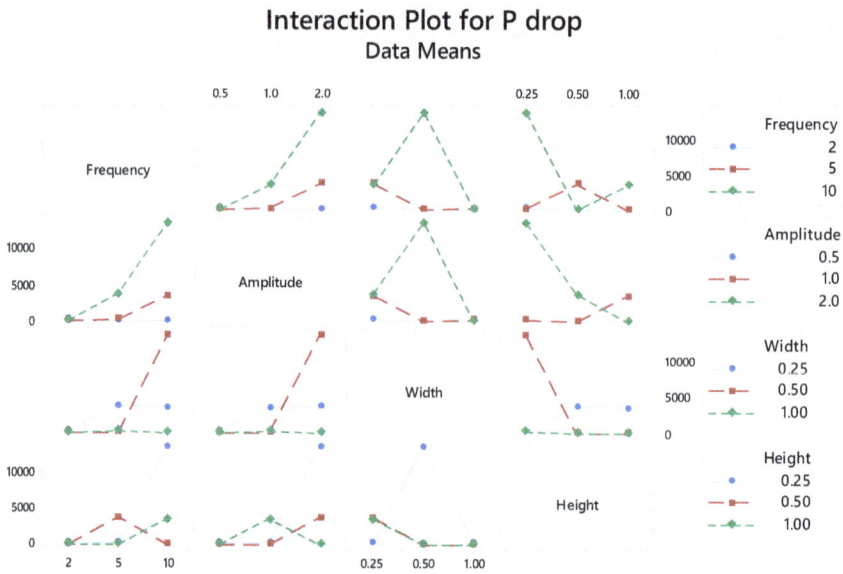

Figure 6. The interactions of various parameters with respect to pumping power.

Now, the combination of optimum factors is evaluated to get the design with the lowest possible pumping power. Table 4 depicts the optimum (minimum) pumping power required. It is noted that, at optimum condition, the pumping power required is 21.75 Pa which is about half than that of design number 3. The level of confidence from Taguchi prediction is observed to be 91.8% which is good enough for engineering design.

4.4. Figure of Merit

So far, we have evaluated the micro-mixer based on mixing performance and pumping power separately. To balance and take into account both effects, we consider the figure of merit, which basically is defined as mixing performance per unit pumping power. Table 3 shows the FoM of OA condition. We note that the highest FoM is achieved by design number 3 due to reasonable mixing performance with the lowest pressure drop requirement; whereas the lowest FoM is seen in design number 9, since the pumping power is very high (about three orders-of-magnitude) compared to design number 3.

Looking further to the sensitivity response of S/N ratio for each parameter in Figure 7, it reveals that the trend is similar to that in pumping power: lower wavy frequency, shorter wavy amplitude, longer microchannel width and longer microchannel height. Turning our attention to the interaction of each factor, Figure 8 depicts a significant interaction between frequency and amplitude with all parameters at different levels, while the interaction between channel width and height is marginal.

Thus, the combination of optimum parameters is evaluated. As can be inferred from Table 4, the optimum FoM is seen to be 3.77×10^{-3} which is higher than that of design number 3 which is 3.56×10^{-3}. The level of confidence from the Taguchi prediction is observed to be 95.8%, which indicates that the Taguchi statistical method is a robust method to select for optimum combination of design parameters in micro-mixers.

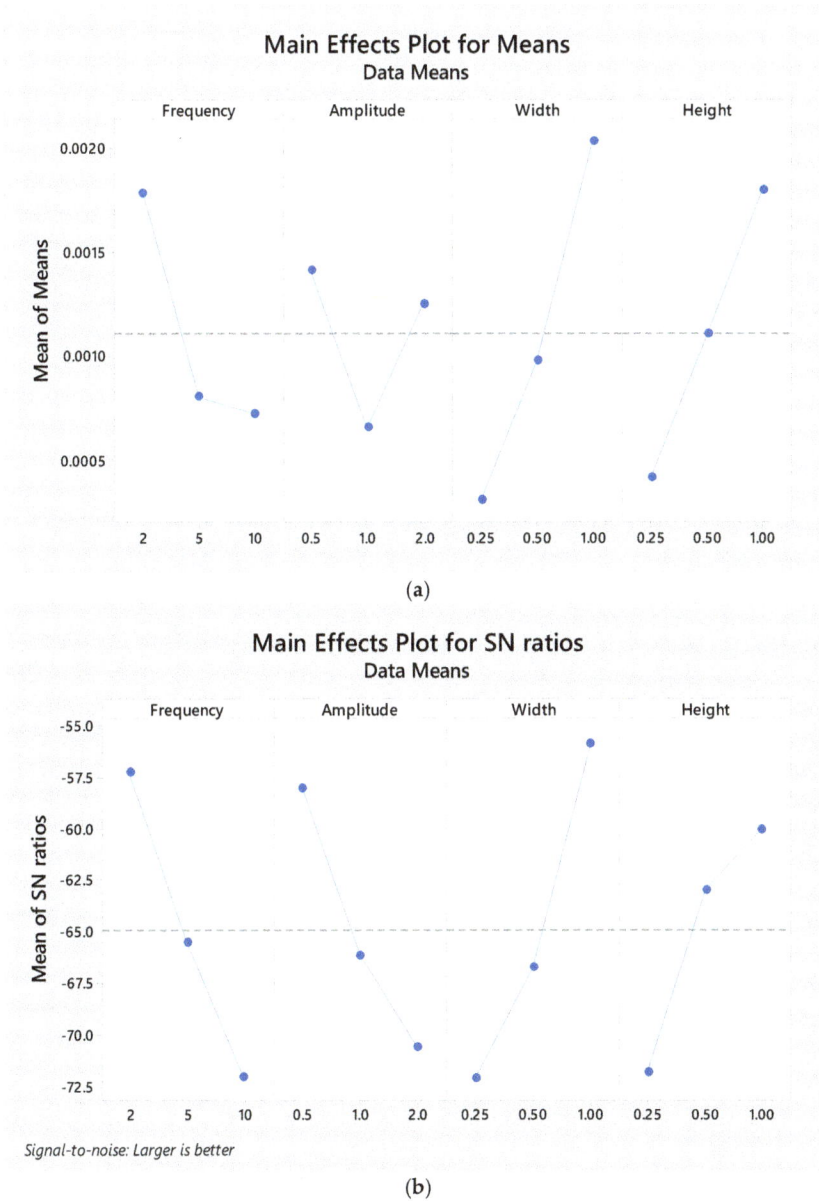

(a)

(b)

Figure 7. Taguchi results of (**a**) Mean and (**b**) S/N response graph for figure of merit (FoM).

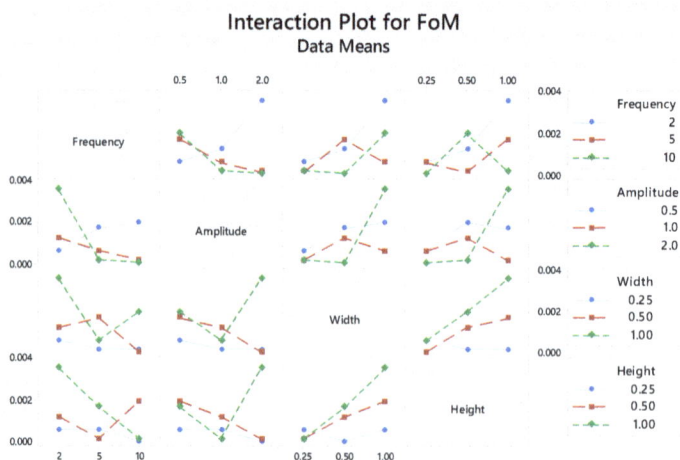

Figure 8. The interactions of various parameters with respect to figure of merit (FoM).

5. Concluding Remarks

A computational study of micro-mixing in microchannel T-junction with wavy structure has been carried out together with the Taguchi statistical method to evaluate the significance of key design parameters with regards to the mixing performance, pumping power and figure of merit. The Taguchi method is found to be robust to determine the optimum combination of design parameters with the maximum relative error of less than 7%.

It has also been shown that, for high value chemical product such as in the pharmaceutical industry, optimization based on mixing index is suggested; the optimum design parameters are high wavy frequency, high wavy amplitude, narrow width and short channel cross-section, while for cheap value and mass production product, for which production cost is important, for example in food industries, optimization based on pumping power is recommended. The optimum design parameters are low frequency, low amplitude, wide and tall channel cross-section. On the other hand, one can also optimize the design based on the figure of merit for middle-value products by taking into account both mixing performance as well as pumping power, for which the parameters are the same as optimization based on pumping power. The results presented herein can aid engineers to determine the best design for micro-mixer performance. Future work will focus on more rigorous optimization procedures in order to alleviate the current limitation of discrete level optimization parameters. Possible coupling of optimization software with CFD software will also be explored.

Acknowledgments: This paper was supported by research funds of Chonbuk National University in 2013.

Author Contributions: N.S., A.P.S., and J.B. conceived and designed the experiments; A.P.S. performed the experiments; N.S. and A.P.S. analyzed the data; A.P.S. contributed analysis tools; N.S., A.P.S., and J.B. wrote the paper.

Conflicts of Interest: The funding sponsors had no role in the design of the study; in the collection, analyses, or interpretation of data; in the writing of the manuscript, and in the decision to publish the results.

References

1. Liu, H.; Li, P.; Lew, J.V. CFD study on flow distribution uniformity in fuel distributors having multiple structural bifurcations of flow channels. *Int. J. Hydrog. Energy* **2010**, *35*, 9186–9198. [CrossRef]
2. Kumar, V.; Paraschivoiu, M.; Nigam, K.D.P. Single-phase fluid flow and mixing in microchannels. *Chem. Eng. Sci.* **2011**, *66*, 1329–1373. [CrossRef]

3. Nguyen, N.T.; Wu, Z. Micromixers—A review. *J. Micromech. Microeng.* **2005**, *15*, R1–R16. [CrossRef]
4. Hessel, V.; Lowe, H.; Schonfeld, F. Micromixers—A review on passive and active mixing principles. *Chem. Eng. Sci.* **2005**, *60*, 2479–5501. [CrossRef]
5. Cai, G.; Xue, L.; Zhang, H.; Lin, J. A review on micromixers. *Micromachines* **2017**, *8*, 274. [CrossRef]
6. Lee, C.Y.; Wang, W.T.; Liu, C.C.; Fu, L.M. Passive mixers in microfluidic systems: A review. *Chem. Eng. J.* **2016**, *288*, 146–160. [CrossRef]
7. Huang, P.H.; Xie, Y.; Ahmed, D.; Rufo, J.; Nama, N.; Chen, Y.; Chan, C.Y.; Huang, T.J. An acoustofluidic micromixer based on oscillating sidewall sharp-edhes. *Lab Chip* **2013**, *13*, 3847. [CrossRef] [PubMed]
8. Nama, N.; Huang, P.H.; Huang, T.J.; Costanzo, F. Investigation of micromixing by acoustically oscillated sharp-edges. *Biomicrofluidics* **2016**, *10*, 024124. [CrossRef] [PubMed]
9. Tseng, W.K.; Lin, J.L.; Sung, W.C.; Chen, S.H.; Lee, G.B. Active micro-mixers using surface acoustic waves on Y-cut 128° LiNbO$_3$. *J. Micromech. Microeng.* **2006**, *16*, 539–548. [CrossRef]
10. Tovar, A.R.; Lee, A.P. Lateral cavity acoustic transducer. *Lab Chip* **2009**, *9*, 41–43. [CrossRef] [PubMed]
11. Tonkovich, A.; Kuhlmann, D.; Rogers, A.; McDaniel, J.; Fitzgerald, S.; Arora, R.; Yuschak, T. Microchannel technology scale-up to commercial capacity. *Chem. Eng. Res. Des.* **2005**, *83*, 634–639. [CrossRef]
12. Lim, W.S.; Choi, H.S.; Ahn, S.Y.; Kim, B.M. Cooling channel design of hot stamping tools for uniform high-strength components in hot stamping process. *Int. J. Adv. Manuf. Technol.* **2014**, *10*, 1189–1203. [CrossRef]
13. Yoon, D.H.; Tanaka, D.; Sekiguchi, T.; Shoji, S. Microfluidic stamping on sheath flow. *Small* **2016**, *24*, 3224–3228. [CrossRef] [PubMed]
14. Bong, H.J.; Lee, J.; Kim, J.H.; Barlat, F.; Lee, M.G. Two-stage forming approach for manufacturing ferritic stainless steel bipolar plates in PEM fuel cell: Experiments and numerical simulations. *Int. J. Hydrog. Energy* **2017**, *42*, 6965–6977. [CrossRef]
15. Choi, J.H.; Oh, C.M.; Jang, J.W. Micro- and nano-patterns fabricated by emboseed microscale stamp with trenched edges. *RSC Adv.* **2017**, *7*, 32058. [CrossRef]
16. Hossain, S.; Lee, I.; Kim, S.M.; Kim, K.Y. A micromixer with two-layer serpentine crossing channels having excellent mixing performance at low Reynolds numbers. *Chem. Eng. J.* **2017**, *327*, 268–277. [CrossRef]
17. Hossain, S.; Kim, K.Y. Mixing analysis in a three-dimensional serpentine split-and-recombine micromixer. *Chem. Eng. Res. Des.* **2015**, *100*, 95–103. [CrossRef]
18. Ahmed, F.; Kim, K.Y. Parametric study of an electroosmotic micromixer with heterogeneous charged surface patches. *Micromachines* **2017**, *8*, 199. [CrossRef]
19. Xie, T.; Xu, C. Numerical and experimental investigations of chaotic mixing behavior in an oscillating feedback micromixer. *Chem. Eng. Sci.* **2017**, *171*, 303–317. [CrossRef]
20. Solehati, N.; Bae, J.; Sasmito, A.P. Numerical investigation of mixing performance in microchannel T-junction with wavy structure. *Comput. Fluids* **2014**, *96*, 10–19. [CrossRef]
21. Solehati, N.; Bae, J.; Sasmito, A.P. Numerical investigation of multi-scale mixing in microchannel T-junction with wavy structure. In Proceedings of the ASME 2012 International Mechanical Engineering Congress and Exposition, Houston, TX, USA, 9–15 November 2012. [CrossRef]
22. Solehati, N.; Bae, J.; Sasmito, A.P. Optimization of operating parameters for liquid-cooled PEM fuel cell stakcs using Taguchi method. *J. Ind. Eng. Chem.* **2012**, *18*, 1039–1050. [CrossRef]
23. Sasmito, A.P.; Kurnia, J.C.; Shamim, T.; Mujumdar, A.S. Optimization of an open-cathode polymer electrolyte fuel cells stack utilizing Taguchi method. *Appl. Energy* **2017**, *185*, 1225–1232. [CrossRef]
24. Barker, T.A. *Quality by Experimental Design*; CRC Press: Boca Raton, FL, USA, 2005.
25. Fowlkes, W.Y.; Creveling, C.M. *Engineering Methods for Robust Product Design Using Taguchi Methods in Technology and Product Development*; Addison-Wesley Publishing Company: Boston, MA, USA, 1995.

micromachines

MDPI

Article

Multi-Objective Optimizations of a Serpentine Micromixer with Crossing Channels at Low and High Reynolds Numbers

Wasim Raza, Sang-Bum Ma and Kwang-Yong Kim *

Department of Mechanical Engineering, Inha University, Incheon 22212, Korea; wasimkr@live.in (W.R.); msb927@inha.edu (S-B.M.)
* Correspondence: kykim@inha.ac.kr; Tel.: +82-32-872-3096; Fax: +82-32-868-1716

Received: 24 January 2018; Accepted: 1 March 2018; Published: 4 March 2018

Abstract: In order to maximize the mixing performance of a micromixer with an integrated three-dimensional serpentine and split-and-recombination configuration, multi-objective optimizations were performed at two different Reynolds numbers, 1 and 120, based on numerical simulation. Numerical analyses of fluid flow and mixing in the micromixer were performed using three-dimensional Navier-Stokes equations and convection-diffusion equation. Three dimensionless design variables that were related to the geometry of the micromixer were selected as design variables for optimization. Mixing index at the exit and pressure drop through the micromixer were employed as two objective functions. A parametric study was carried out to explore the effects of the design variables on the objective functions. Latin hypercube sampling method as a design-of-experiment technique has been used to select design points in the design space. Surrogate modeling of the objective functions was performed by using radial basis neural network. Concave Pareto-optimal curves comprising of Pareto-optimal solutions that represents the trade-off between the objective functions were obtained using a multi-objective genetic algorithm at $Re = 1$ and 120. Through the optimizations, maximum enhancements of 18.8% and 6.0% in mixing index were achieved at $Re = 1$ and 120, respectively.

Keywords: micromixer; multi-objective optimization; Reynolds number; Navier-Stokes equations; surrogate modeling; pressure drop

1. Introduction

Microfluidics is related to an expeditiously emerging technology enabling manipulation and control of minute volumes of fluids with high accuracy in a miniaturized system for various fluidic functions, such as transporting, metering, valving, mixing, reacting, and separating [1,2]. Micromixer is an integral component of the microfluidic systems that have promising impact in the fields of biomedical diagnostics, drug development, and chemical industry [3,4]. Efficient mixing of liquid samples is a challenging task for successful operation of different processes in the microfluidic systems. The flow nature in the microfluidic systems is laminar, due to low Reynolds number. Thus, the mixing of fluid species depends mainly on mass diffusion in the absence of turbulence. However, the diffusion-dependent mixing is relatively slow and ineffective. In order to enhance the mixing performance, numerous methods have been proposed during the last two decades [3,5–7].

Depending upon the working principle, micromixers are categorized either as a passive or as an active type. Active micromixers employ external energy sources, such as electrokinetic, ultrasonic vibration and magnetic field to generate flow perturbations inside the microchannel. Although active micromixers generally show excellent mixing capability and control during the mixing, high fabrication cost, and difficulty in integration with microfluidic systems make them less practical. In contrast, passive micromixers that enhance mixing by modifying the microchannel geometry, are being widely developed due to the advantages of simple fabrication and easy integration with the microfluidic systems [3,5–8].

Over the last two decades, many researchers have proposed different microchannel designs to enhance the mixing performance of passive micromixers. In general, passive micromixers, depending upon the mechanism of mixing, are classified as lamination-based [9–13] or chaotic advection-based [14–17]. Kim et al. [9] proposed a serpentine laminating micromixer that combines the mixing mechanisms of split-and-recombination (SAR) using successive arrangement of F-shaped mixing units in two layers and chaotic advection through a three-dimensional (3D) serpentine microchannel path. Tofteberg et al. [10] developed a lamination micromixer incorporating a sequence of SAR process with patterning of the channel bottom with grooves. Nimafar et al. [12] proposed an H-micromixer based on SAR process, and compared its mixing performance with those of T- and O-micromixer in a Reynolds number range of 0.08–4.16. The results showed that the H-micromixer achieved 98% mixing at Re = 0.083 due to SAR process, which was much higher than those of the other two micromixers.

Liu et al. [14] proposed a 3D serpentine micromixer with C-shaped repeating units, and experimentally demonstrated its high mixing performance for Reynolds numbers larger than 25. Stroock et al. [15] proposed a micromixer with patterned grooves on the channel bottom to induce chaotic mixing at low Reynolds numbers. Xia et al. [16] proposed two chaotic micromixers consisting of double-layer crossing channels. Their numerical and experimental results demonstrated that chaotic advection that is generated through continuous stretching and folding along with splitting and recombination, even at low Reynolds numbers (Re < 0.2), resulted in high mixing performance. The et al. [18] proposed a shifted trapezoidal blade micromixer that combined several mixing principles, i.e., vortices, transversal flows, and chaotic advection, to attain a stable mixing efficiency of larger than 80% in a wide Reynolds number range of 0.5–100.

Numerical optimization techniques coupled with computational fluid dynamics (CFD) analysis based on 3D Navier-Stokes equations have been widely used as an efficient tool for design of micromixers [19–23]. Ansari and Kim [19,20] optimized geometric parameters of a staggered herringbone groove micromixer (SHM) with grooves at the bottom wall by using radial basis neural network (RBNN) and response surface approximation (RSA) as surrogate models. These optimizations were carried out using mixing index as a single objective function with two or three design variables. Kim et al. [21] performed an optimization of microscale vortex generators in a micromixer using an advanced RSA by considering four geometric design variables. It was found that a mixing uniformity of larger than 95% was obtained within a channel length of 1344 µm with an optimized microchannel configuration. Cortes-Quiroz et al. [22] applied a multi-objective genetic algorithm (MOGA) for shape optimization of a SHM at Re = 1 and 10, by integrating CFD calculations with a surrogate model. In this optimization, a Pareto-optimal front of trade-offs between two objective functions, i.e., mixing index at the exit and pressure drop, was generated by MOGA. Afzal and Kim [23] performed a multi-objective optimization of a SHM with three objective functions, i.e., mixing index at the exit, friction factor, and mixing sensitivity. Hossain and Kim [24] optimized a micromixer with two-layer serpentine crossing channels at Reynolds numbers of 0.2 and 40 by using CFD, surrogate model and MOGA. Kriging metamodel was used as a surrogate model, while mixing index at the exit and pressure drop were used as two objective functions to generate Pareto-optimal front.

In a recent study, Raza et al. [25] proposed a 3D serpentine SAR micromixer with OX-shaped mixing units, and analyzed the flow structure and mixing performance numerically in a wide Reynolds number range of 0.1–200. The proposed micromixer with five mixing units showed excellent mixing performance over the entire range of Reynolds number, through stretching and folding of fluid interface at the crossing channel intersection nodes at low Reynolds numbers and chaotic motion due to the 3D serpentine path at high Reynolds numbers. Especially, the enhancement of mixing performance in low-Re range is noticeable as compared to previous micromixers.

In the present work, multi-objective optimizations have been performed to further enhance the performance of the micromixer proposed by Raza et al. [25], at both low and high Reynolds numbers. Mixing index at the exit of the micromixer and pressure drop though the micromixer were employed as two objective functions. The Pareto-optimal fronts compromising these two objectives were generated using RBNN surrogate model and MOGA. The optimizations were carried out with the two objective functions at the two different Reynolds numbers, $Re = 1$ and 120. The corresponding Peclet numbers are 1×10^4 and 120×10^4, respectively.

2. Micromixer Geometry

The micromixer configuration, as proposed by Raza et al. [25], has been used for the optimization. The micromixer consists of two repeated OX-shaped mixing units of 3D serpentine SAR structure as shown in Figure 1. The number of mixing units was reduced from five of previous work [25] to two in this work. Chaotic advection through stretching and folding of fluid interface at the X-junction, where the fluids in both the crossing channels are exchanged, enhances mixing at low Reynolds numbers. On the other hand, 3D serpentine path that was created by O-structure promotes mixing by producing chaotic advection at high Reynolds numbers, as explained in the previous study [25].

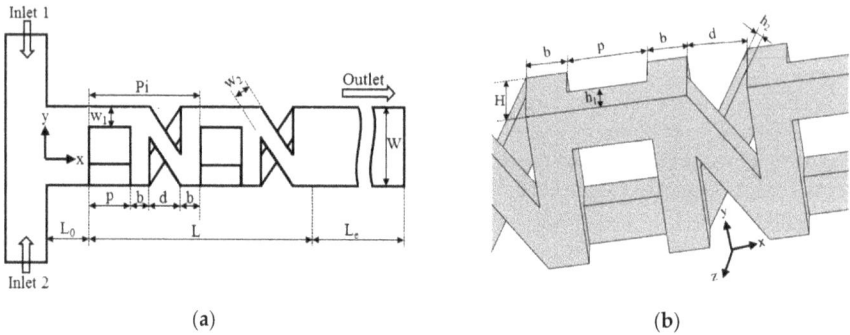

(a)   (b)

Figure 1. Geometry of micromixer with OX- shaped units: (a) planar view [25]; and, (b) three-dimensional (3D) view.

Values of the geometric parameters for the reference micromixer are listed in Table 1. The traditional method of stacking the polydimethylsiloxane (PDMS) layers can be used to fabricate the proposed micromixer. As described in the previous works [17,26], the micro-molding technology, i.e., soft lithography technique using SU-8 master molds, can be employed to fabricate these PDMS layers.

Table 1. Geometric parameters and their values.

Geometric Parameter	Value (In μm)
Length of initial part of main channel, L_0	100
Exit channel length, L_e	1500
Total length, L_t	2150
Pitch length, Pi	275
Width of main channel, W	200
Width of O-structure, w_1	50
Width of X-structure, w_2	50
Distance between O-structures, d	75
Depth of horizontal portion of O-structure, h_1	50
Depth of X-structure, h_2	50
Total depth of main channel, H	100

3. Numerical Analysis

The analyses of fluid flow and mixing inside the micromixer were performed numerically using a commercial CFD code, ANSYS CFX 15.0 (ANSYS, Inc., Canonsburg, PA, USA) [27], based on finite volume approximations. The numerical analysis of steady, incompressible, 3D laminar flow was carried out by solving the continuity and Navier-Stokes equations:

$$\nabla \cdot \vec{V} = 0 \tag{1}$$

$$\left(\vec{V} \cdot \nabla\right)\vec{V} = -\frac{1}{\rho}\nabla P + \nu\nabla^2\vec{V} \tag{2}$$

where the symbols ν, ρ, and $\vec{V}$ denote the fluid kinematic viscosity, density, and velocity, respectively. A solution of dye in water and water at a temperature of 20 °C were considered as two working fluids. To model the mass transport of the fluids having constant viscosity and density in the mixing process, a scalar transport equation of advection-diffusion type [28] was used, as follows:

$$\left(\vec{V} \cdot \nabla\right)C = \alpha\nabla^2 C \tag{3}$$

where the symbols α and C denote the diffusivity coefficient and dye concentration, respectively. In order to model mixing of two fluids, the scalar transport equation has been used and validated previously for different micromixers [29,30].

The computational domain was discretized by a combination of tetrahedral and hexahedral elements to reduce the total number of computational nodes. The mixing units were meshed with tetrahedral element due to their complex geometry, while the remaining part in the microchannel was meshed with hexahedral elements.

Uniform velocity profiles were assigned at the inlets, while the atmospheric pressure was used at the outlet. At the walls, no-slip condition was used. A solution of dye in water (mass fraction equal to 1) and water (mass fraction equal to 0) were introduced at the inlet 1 and inlet 2, respectively. The diffusivity coefficient value of the water-dye mixture was 1.0×10^{-10} m²/s. The values of density and dynamic viscosity of water (and also water-dye mixture) were 1000 kg/m³ and 1.0×10^{-3} kg·m⁻¹·s⁻¹, respectively [31]. Truncation errors that were associated with numerical discretization for the advection terms in the governing differential equations give rise to numerical diffusion. The extent of numerical diffusion depends on the numerical scheme used. Numerical diffusion can be decreased by using higher order approximation schemes, such as second-order upwind and third order QUICK [32] scheme instead of first-order upwind scheme [33]. A high-resolution scheme of second-order approximation was applied for discretization of advection terms. The solution convergence criterion was set as root-mean-square residual value of less than 1.0×10^{-6}.

A statistical method [34] based on concept of intensity of segregation was used to define mixing index. The mixing index at a plane perpendicular to streamwise direction is represented as follows:

$$M = 1 - \frac{\sigma}{\sigma_{max}} \tag{4}$$

where σ and σ_{max} are the standard deviation of the concentration at the cross-sectional plane, and maximum standard deviation over the entire data range, respectively.

The standard deviation of dye mass fraction at the cross-sectional plane can be written as:

$$\sigma = \sqrt{\frac{1}{N}\sum_{i=1}^{N}(c_i - \bar{c}_m)^2} \tag{5}$$

where c_i, $\bar{c}_m$, and N are the mass fraction at sampling point i, the optimal mixing mass fraction, and the number of sampling points on the plane, respectively. A mixing index value of 0 indicates completely unmixed fluids, while a value of 1 indicates completely mixed fluids. Mixing index at the exit (M_o) was defined as the mixing index 700 μm downstream of the main channel starting position.

The Reynolds Number was defined using hydraulic diameter (D_h) of the main channel as follows:

$$Re = \frac{\rho V D_h}{\mu} \tag{6}$$

where ρ, μ, and V denote the density of water, dynamic viscosity of water, and fluid average velocity, respectively.

Hydraulic diameter was defined as follows:

$$D_h = \frac{2 \times W \times H}{(W + H)} \tag{7}$$

where W and H denote the width and depth of the microchannel.

4. Design Variables and Objective Functions

For the optimization, the ratios of distance between O-structures to pitch (d/Pi), crossing channel width to pitch (w_2/Pi), and total depth of the micromixer to main channel width (H/W), as shown in Figure 1, were selected as design variables among various geometric parameters through a preliminary parametric study. Effects of these design variables on mixing index at the exit and pressure drop through the micromixer at $Re = 1$ for the reference micromixer, are shown in Figure 2. For all of the design variables, the mixing index shows maximum values in the tested ranges of the variables. Pressure drop increases with increase in d/Pi, while it decreases with increase in w_2/Pi and H/W in the tested ranges. Design ranges for the design variables are summarized in Table 2.

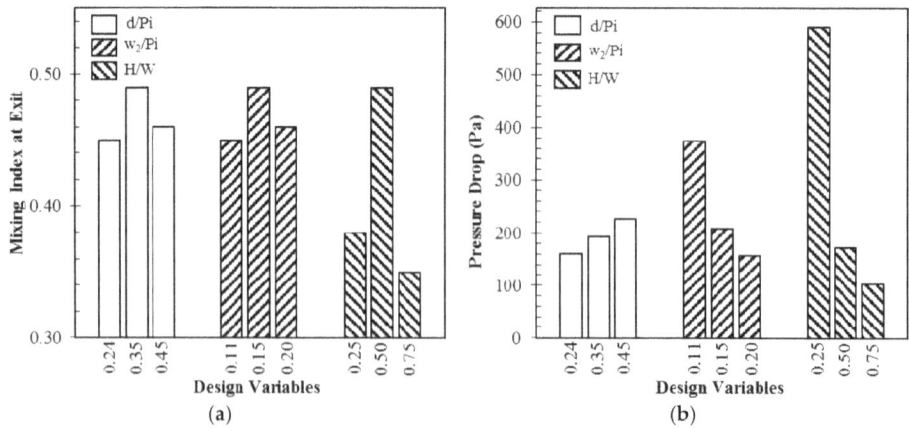

Figure 2. Effects of design variables on objective functions at $Re = 1$: (a) mixing index at the exit; and, (b) pressure drop.

In this work, mixing index at the exit and pressure drop in the micromixer were used to define two objective functions. The pressure drop that affects the power required to drive the fluids through the micromixer, was calculated as the difference in area-weighted average of the total pressure between the planes at the main channel inlet and exit. To make the optimization problem be defined as minimization of the objective functions, one of the objective functions (F_M) was taken as the negative of mixing index, and the other objective function ($F_{\Delta P}$) was defined as pressure drop. Two pairs of objective functions were selected as follows: mixing index at $Re = 1$ ($F_{M \text{ at } Re = 1}$)—pressure drop at $Re = 1$ ($F_{\Delta P \text{ at } Re = 1}$) and mixing index at $Re = 120$ ($F_{M \text{ at } Re = 120}$)—pressure drop at $Re = 120$ ($F_{\Delta P \text{ at } Re = 120}$). Values of the objective functions for the reference design are listed in Table 3.

Table 2. Design variables and their ranges.

Design Variables	Lower Limit	Upper Limit
d/Pi	0.236	0.455
w_2/Pi	0.109	0.200
H/W	0.250	0.750

Table 3. Reference design and its objective functions values.

Reference Design						
Design Variables			Objective Functions			
d/Pi	w_2/Pi	H/W	Mixing Index at exit, M_0		Pressure drop, ΔP (kPa)	
			$Re = 1$	$Re = 120$	$Re = 1$	$Re = 120$
0.272	0.182	0.500	0.489	0.883	0.171	69.741

5. Surrogate Model and Multi-Objective Optimization

Procedure of multi-objective optimization used in the present work is represented in Figure 3. The first step is selecting design variables and their ranges through a previous parametric study, and the objective functions considering the design goals. The next step is to build the design space using the design of experiments (DOE). In the present work, Latin Hypercube Sampling (LHS) [35] is used as DOE. LHS is an effective sampling technique that uses an $m \times n$ simulation matrix where m is the number of levels (sampling points) to be examined and n is the number of design parameters. Each of the n columns of the matrix containing the levels, $1, 2, \ldots, m$, is randomly paired to form a Latin hypercube. This approach produces random sample points, ensuring that all of the portions of the design space are represented. Objective function values are calculated by Navier-Stokes analysis at these design points. A surrogate model is constructed to approximate the objective functions based on these objective function values, and MOGA is used to explore global Pareto-optimal solutions in the design space. If the global optimum solution exists in the design space and a termination criterion is satisfied, the multi-objective optimization procedure terminates.

The multi objective optimization problem was formulated, as follows:

$$\text{Minimization: } F(x) = [F_1(x), F_2(x), F_3(x) \ldots F_n(x)]$$

$$\text{Design variable bound: } LB \leq x \leq UB, x \in R$$

where $F(x)$ is an objective function, x is a vector of design variables, and LB and UB denote the vectors of the lower and upper bounds, respectively.

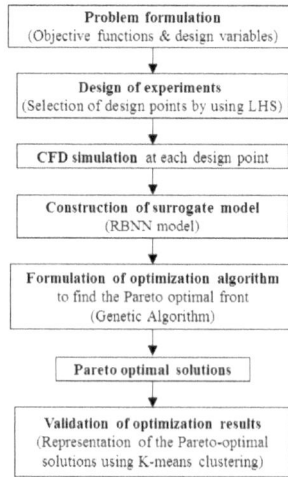

Figure 3. Multi-objective optimization procedure.

In order to obtain LHS design points, MATLAB (MathWorks, Inc., Natick, MA, USA) function '*lhsdesign*' [36] was used with the criterion '*maxmin*' (maximize minimum distance between adjacent design point). As a result, uniformly distributed 27 design points were selected for the three design variables in order to construct a surrogate model. The objective function values calculated at these design points by Navier-Stokes analysis are listed in Table 4.

Table 4. Design variables and objective function values at Latin hypercube sampling (LHS) design points.

Design Point	Design Variables			Objective Functions (*Re* = 1)		Objective Functions (*Re* = 120)	
	d/Pi	*w₂/Pi*	*H/W*	Mixing Index, M_o	Pressure Drop, ΔP (kPa)	Mixing Index, M_o	Pressure Drop, ΔP (kPa)
1	0.377	0.162	0.635	0.410	0.962	0.171	61.976
2	0.418	0.197	0.673	0.400	0.926	0.136	48.443
3	0.369	0.172	0.442	0.504	0.962	0.254	91.358
4	0.288	0.169	0.423	0.539	0.929	0.248	98.528
5	0.442	0.141	0.577	0.404	0.854	0.261	89.404
6	0.272	0.131	0.731	0.374	0.977	0.171	69.802
7	0.385	0.113	0.500	0.419	0.750	0.422	153.596
8	0.434	0.148	0.385	0.488	0.795	0.422	143.310
9	0.321	0.155	0.712	0.391	0.915	0.123	46.962
10	0.329	0.183	0.692	0.382	0.930	0.232	86.091
11	0.345	0.127	0.654	0.371	0.847	0.334	114.287
12	0.426	0.117	0.615	0.468	0.920	0.350	151.380
13	0.280	0.120	0.462	0.474	0.972	0.234	88.623
14	0.345	0.155	0.500	0.472	0.935	0.192	68.240
15	0.402	0.186	0.519	0.472	0.935	0.192	68.240
16	0.256	0.190	0.365	0.459	0.862	0.266	105.153
17	0.337	0.179	0.327	0.580	0.914	0.391	136.554
18	0.353	0.124	0.288	0.495	0.831	0.810	300.789
19	0.361	0.152	0.269	0.572	0.852	0.715	247.365
20	0.248	0.158	0.596	0.450	0.885	0.150	65.947
21	0.297	0.193	0.481	0.499	0.888	0.178	68.949
22	0.410	0.176	0.250	0.541	0.818	0.754	229.751
23	0.240	0.165	0.404	0.543	0.915	0.248	108.789
24	0.393	0.138	0.558	0.420	0.891	0.261	93.075
25	0.313	0.134	0.538	0.448	0.933	0.250	99.692
26	0.264	0.110	0.308	0.493	0.941	0.736	342.159
27	0.305	0.145	0.346	0.534	0.911	0.432	173.488

A RBNN model [37] was used to construct surrogate models of the objective functions. The RBNN model contains a two-layered network comprised of a hidden layer of radial neuron and an output layer of linear neuron. The hidden layer executes a nonlinear alteration of the input space to a middle space by using a set of radial basis elements. The output layer then implements a linear combiner to yield the desired targets. The linear model $g(x)$ for the function can be expressed as a linear combination of a set of M radially-symmetric functions, as follows:

$$g(x) = \sum_{j=1}^{M} w_j \Phi_j \tag{8}$$

where w_j is weights and Φ_j are radial basis functions. Benefit of this surrogate modeling is an ability to reduce computational time owing to the linear nature of the radial basis functions. In the present work, the RBNN function, *newrb* [36] was used to construct the models. The network training was performed by varying spread constant (SC) to adjust the cross-validation error. In this study, SC_1 is related to mixing index, and SC_2 is related to pressure drop. The SC values of each objective function were selected through a k-fold cross-validation [38] error test in which the errors were minimum at $SC_1 = 0.9$ and $SC_2 = 0.5$, as shown in Figure 4. The k-fold cross-validation is a validation method to estimate how the results of a statistical analysis generalize to an independent data set.

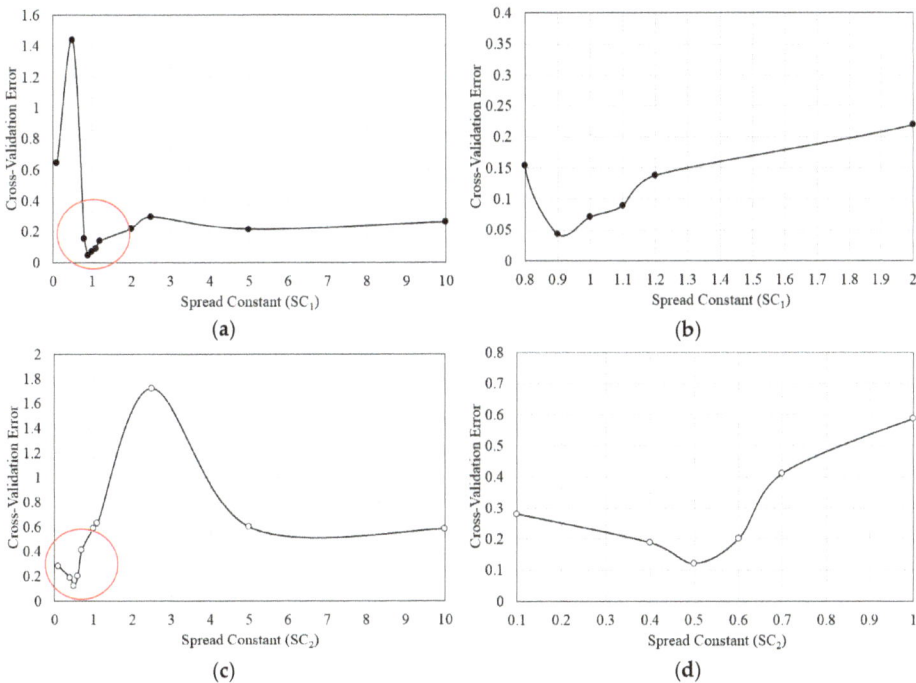

Figure 4. Cross-validation error vs. spread constant (SC) for the objective functions: (**a**) Mixing index (SC_1); (**b**) Zoomed in view of encircled area in Figure 4a; (**c**) Pressure drop (SC_2); (**d**) Zoomed in view of encircled area in Figure 4c.

MOGA was used to find optimal solution on the constructed surrogate model using the MATLAB optimization tool box [36]. MOGA is a randomized global search method that solves functions by imitating progressions observed from natural evolution [37]. Based on the survival and growth of the fittest, MOGA finds new and improved results repeatedly. MOGA describes an initial population of individuals, which represent a part of the solution to the functions. Before the search begins, a series of chromosomes is randomly selected from the design space to obtain the initial population. Through computations, the individuals are selected in a competitive way, based on their fitness functions as measured by each specific objective function. The genetic search operators (i.e., "selection", "mutation", and "crossover") were applied to obtain a next generation of chromosomes, for which the predicted quality of all the chromosomes is better than that of the previous generation. This process is repeated until the termination criterion, which is function tolerance, is met. The following parameters were used: population size = 400, cross over fraction = 0.7, generations = 800, and function tolerance =10^{-8}.

6. Results and Discussion

A grid-dependency test was performed to find out an optimal number of nodes for the spatial discretization of computational space. Four different grid systems with 4.78×10^5 to 1.43×10^6 nodes were tested for development of mixing index along the channel length at two different Reynolds numbers (*Re* = 1 and 120), as shown in Figure 5. The mixing indices were calculated on the planes perpendicular to the axial direction at four different locations (i.e., start of the main channel, two successive intersection nodes of the crossing channels in X-structure, and the exit). Almost similar profiles of the mixing index development are observed for grid systems with 1.21×10^6 and 1.43×10^6 nodes at both *Re* = 1 and 120. Hence, grid system with 1.21×10^6 nodes was selected as an optimal grid, commonly at *Re* = 1 and 120.

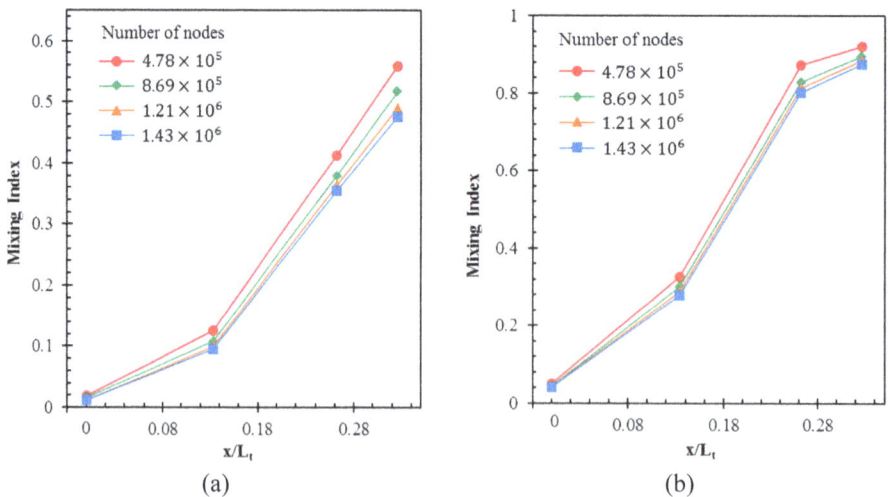

Figure 5. Grid-dependency tests for mixing index development along axial length of the reference design: (**a**) *Re* = 1; and, (**b**) *Re* = 120.

The present numerical model was validated in the previous study [25] by both qualitative and quantitative comparisons of the numerical results with the experimental results of Hossain et al. [39], as shown in Figure 6. The optical image of fluid mixing in the first mixing unit at *Re* = 60 is compared with the numerical result of dye mass fraction distribution on *x-y* plane located halfway along the channel depth, as shown in Figure 6a. The numerical values of mixing indices at the exit of the micromixer in a Reynolds number range of 0.2–120 are also compared with the experimental data, as shown in Figure 6b. The numerically predicted mixing indices are slightly higher than the experimental data over the whole range. The uncertainties that are involved in the experimental procedures, such as capturing and analyzing experimental image, geometrical variation in fabrication, and wall roughness, can be attributed as the causes for the differences in the mixing indices, as shown in Figure 6b. However, the qualitative and quantitative comparisons between the numerical and experimental results, show acceptable agreements.

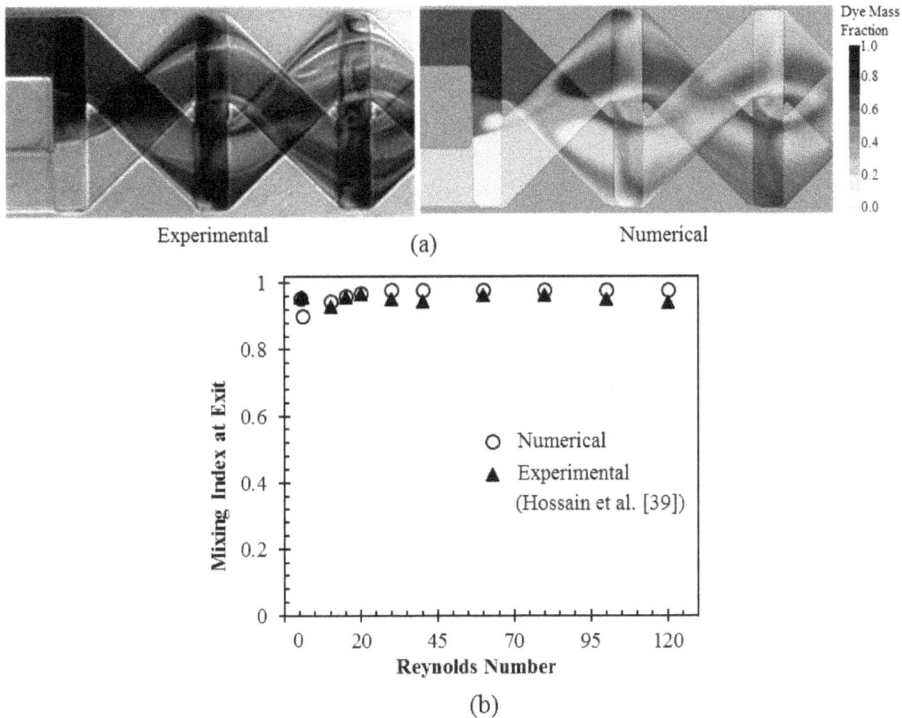

Figure 6. Validation of numerical results compared with experimental data [25]: (**a**) qualitative comparison of dye mass fraction distribution on *x-y* plane located halfway along the channel depth between experiment and numerical analysis at *Re* = 60; and, (**b**) quantitative comparison of mixing indices at the exit for different Reynolds numbers between experiment and numerical analysis.

Figure 7 shows effect of number of the mixing units on the mixing performance through a quantitative comparison among the mixing indices at the exits of the micromixers with two, three, and five mixing units at different Reynolds numbers. It is observed that the micromixer with five mixing units exhibits almost complete mixing in most of the Reynolds number range, while the micromixers with two and three mixing units show largely reduced mixing index values over the whole *Re* range. It is also observed that all the micromixers show minimum mixing indices at *Re* = 1. Hence, the micromixer with two mixing units was selected in the present work to have more space for enhancement of the mixing performance by optimization.

Effect of Reynolds number on development of mixing index along the axial length of the micromixer (reference design) is shown in Figure 8. The mixing indices were calculated on four *y-z* planes (at the start of the main channel, the two successive intersection nodes of the crossing channels in X-structure, and the exit) for Reynolds numbers, 1, 20, 40, 80, and 120. The results show that developing rate of mixing index depends on Reynolds number, as shown in Figure 8. The rate of development of mixing index is found to be highest between the first and second intersection nodes at *Re* = 40.

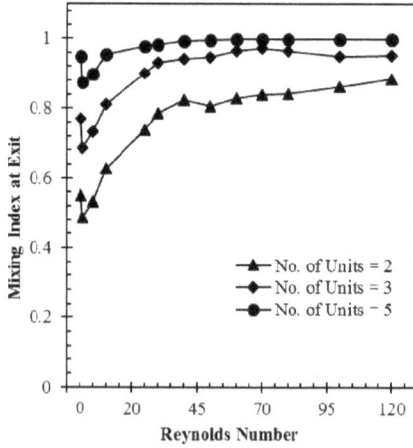

Figure 7. Variations of mixing index at the exit with Reynolds number for three different numbers of mixing units.

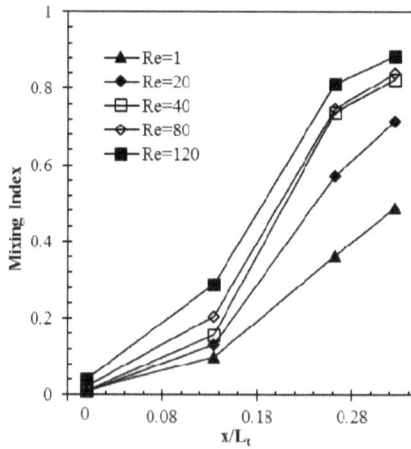

Figure 8. Developments of mixing index along the axial length of micromixer (reference design) for different Reynolds numbers.

Following the procedure outlined in Figure 3, Pareto-optimal fronts presenting the optimum trade-off between the two conflicting objectives are plotted, as shown in Figure 9. Pareto-optimal fronts, termed as POF-1 and POF-2 in the Figure 9a,b, respectively, represent two pairs of the objective functions at two different Reynolds numbers: $F_{M \text{ at } Re = 1} - F_{\Delta P \text{ at } Re = 1}$ and $F_{M \text{ at } Re = 120} - F_{\Delta P \text{ at } Re = 120}$. Concave shape of the Pareto-optimal fronts indicates that an improvement in mixing index occurs with simultaneous increase in pressure drop. Selection of Pareto-optimal solution depends upon the choice of the designer, since each solution is a global Pareto-optimal solution and none of these Pareto-optimal solutions is superior to the others for both objective functions.

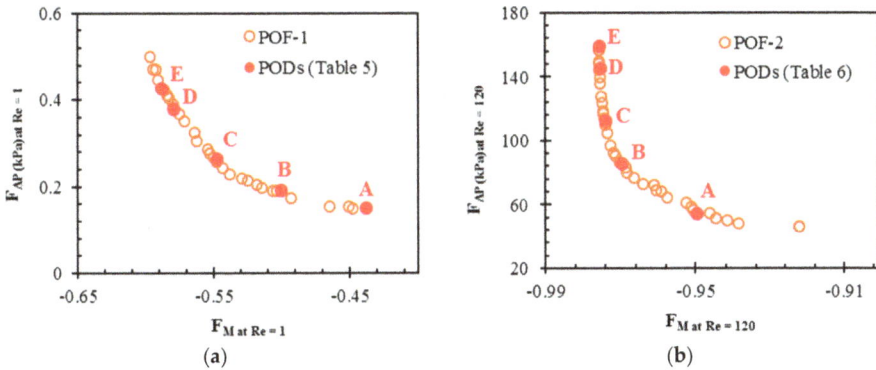

Figure 9. Pareto-optimal fronts: (**a**) mixing index at $Re = 1$ vs. pressure drop at $Re = 1$; and (**b**) mixing index at $Re = 120$ vs. pressure drop at $Re = 120$.

In order to analyze the Pareto-optimal solutions, five representative Pareto-optimal designs (PODs) were selected by K-means clustering on each Pareto-optimal front, as shown in Figure 9. PODs A and E at extreme ends of each Pareto-optimal front represent pressure drop-oriented and mixing index-oriented designs, respectively. Accomplishment of one objective function leads to forfeit of the other objective function. At $Re = 1$, as compared to POD A on POF-1, the mixing index-oriented design, POD E shows a relative enhancement of 34.5 % in the mixing index at the exit. The pressure drop-oriented design, POD A shows a 64.5 % reduction in pressure drop as compared to POD E. Similarly, at $Re = 120$, POD E shows a relative enhancement of 2.8 % in mixing index at the exit as compared to POD A on POF-2, and POD A shows a 65.7% reduction in pressure drop as compared to POD E. These results reveal that the relative enhancement of mixing index at the exit is much more pronounced at $Re = 1$ as compared to $Re = 120$. However, the relative percentage reductions in pressure drop are not much different at both the Reynolds numbers. The higher relative percentage changes in pressure drop than those in mixing index at the exit indicate that mixing index is less sensitive to the design variables as compared to pressure drop.

The results at the representative PODs for POF-1 and POF-2 are listed in Tables 5 and 6, respectively. It is observed that mixing index at the exit and pressure drop are most sensitive to the design variable, H/W, while w_2/Pi remains nearly invariant on POF-1 with the exception of POD A. A high value of mixing index is observed for d/Pi, w_2/Pi, and H/W values close to middle of the range, upper bound, and lower bound, respectively. In case of POF-2, the objective functions become more sensitive to the design variables when compared to POF-1. The results of numerical analysis at the PODs shows that maximum enhancements of 18.8% (POD E on POF-1) and 6.0% (POD E on POF-2) in mixing index at the exit are achieved as compared to the reference design at $Re = 1$ and 120, respectively, by the optimization. Maximum reductions of 5.8% (POD A on POF-1) and 11.1% (POD A on POF-2) in pressure drop are obtained as compared to the reference design at $Re = 1$ and 120, respectively. The surrogate model predictions of the objective functions values are also compared with the numerical results at the same designs in Tables 5 and 6. The relative errors of the surrogate predictions for mixing index at the exit and pressure drop are less than 2% and 10%, respectively, at $Re = 1$, as shown in Table 5. However, these relative errors are increased at $Re = 120$ to less than 5% and 22%, respectively, as shown in Table 6.

Table 5. Results of optimization at representative Pareto-optimal designs (PODs) for $Re = 1$.

Selected PODs	Design Variables			Surrogate Prediction		Numerical Analysis		% Error	
	d/Pi	w_2/Pi	H/W	M_o	ΔP (kPa)	M_o	ΔP (kPa)	M_o	ΔP
A	0.308	0.163	0.602	0.438	0.152	0.439	0.161	−0.23	−5.59
B	0.296	0.171	0.485	0.500	0.190	0.498	0.199	0.40	−4.52
C	0.318	0.175	0.404	0.548	0.265	0.551	0.269	−0.54	−1.49
D	0.327	0.175	0.338	0.580	0.378	0.587	0.373	−1.19	1.34
E	0.330	0.176	0.328	0.589	0.428	0.581	0.391	1.38	9.46

Table 6. Results of optimization at representative PODs for $Re = 120$.

Selected PODs	Design Variables			Surrogate Prediction		Numerical Analysis		% Error	
	d/Pi	w_2/Pi	H/W	M_o	ΔP (kPa)	M_o	ΔP (kPa)	M_o	ΔP
A	0.329	0.161	0.628	0.950	54.544	0.957	61.966	−0.73	−11.98
B	0.272	0.132	0.583	0.970	85.442	0.951	92.696	2.00	−7.83
C	0.254	0.123	0.472	0.974	112.168	0.936	142.565	4.06	−21.32
D	0.245	0.120	0.464	0.976	145.752	0.948	154.631	2.95	−5.74
E	0.241	0.120	0.446	0.976	158.992	0.936	166.768	4.27	−4.66

For qualitative comparison of mixings at different PODs on POF-1 ($Re = 1$) and POF-2 ($Re = 120$), dye mass fraction contours are plotted on y-z planes at the beginning of the main channel ($x/L_t = 0$), two successive intersection nodes of the crossing channel in X-structure ($x/L_t = 0.13$ and 0.26), and the exit ($x/L_t = 0.32$), respectively, as shown in Figure 10. Two Pareto-optimal designs, i.e. POD A and POD E at the extreme ends of each Pareto-optimal front are selected for the comparison. Number of two-fluid interfaces increases along the channel length for both the PODs. However, there is a distinct difference in the dye mass fraction distribution at the exit of the micromixer between POD A and POD E on POF-1. A more uniform dye mass fraction distribution is observed at the exit for POD E as compared to POD A on POF-1, as shown in Figure 10a, whereas almost similar distributions are observed for POD A and POD E on POF-2 (Figure 10b). This is related to the fact that the change in mixing index along POF-1 (0.439–0.581 in Table 5) is much larger than that along POF-2 (0.936–0.957 in Table 6).

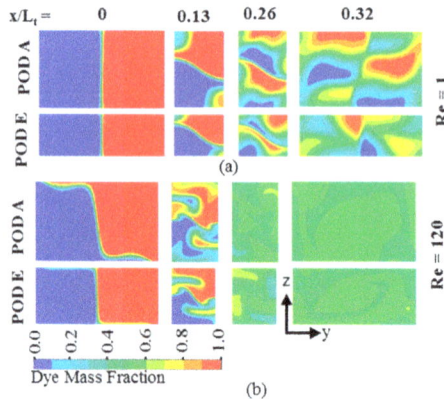

Figure 10. Dye mass fraction distributions on y-z planes for POD A and POD E: (**a**) Pareto-optimal fronts (POF)-1 ($Re = 1$); and (**b**) POF-2 ($Re = 120$).

In order to analyze the mixing mechanism, velocity vectors superimposed on the dye mass fraction contours are plotted for POD A and POD E on POF-1 ($Re = 1$), in Figure 11. It is observed that a saddle-shaped flow structure, which promotes chaotic advection [25], is formed at each intersection node (x/L_t = 0.13 and 0.26) of the crossing microchannels in both of the Pareto-optimal designs, POD A and POD E. However, there is a difference in the fraction of total depth of the microchannel cross-section covered by saddle-shaped pattern between the two designs. In case of POD A, the saddle-shaped flow structure exists only in the middle height of the *y*-*z* plane, as shown in Figure 11a. Whereas, in case of POD E, the saddle-shaped flow pattern covers more height of the *y*-*z* plane, as shown in Figure 11b. This results in more uniform mass fraction distribution (Figure 11b) and consequently, higher mixing index in POD E, as compared to POD A.

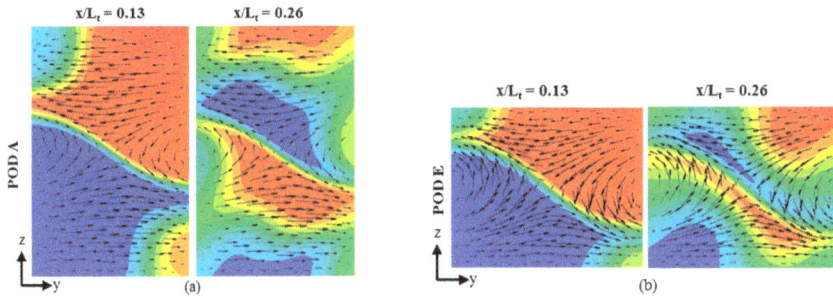

Figure 11. Velocity vectors and concentration contours on *y*-*z* planes at first (x/L_t = 0.13) and second (x/L_t=0.26) X-structure intersection nodes: (**a**) POD A; and, (**b**) POD E, on POF-1.

In order to have an idea about the mixing performance of a Pareto-optimal design at Reynolds numbers other than the Reynolds number at which it was obtained, mixing performances of PODs E on POF-1 and POF-2 are compared with the reference design at Reynolds numbers, 1, 30, 60, 90, and 120, as shown in Figure 12. It is observed that POD E on POF-1 shows values of mixing index larger than those of the reference design except at intermediate Reynolds number of 30, but the mixing index values are smaller than those of POD E on POF-2 except at *Re* = 1. POD E on POF-2 shows improvements in the mixing index at Reynolds numbers larger than 60 when compared to the other designs.

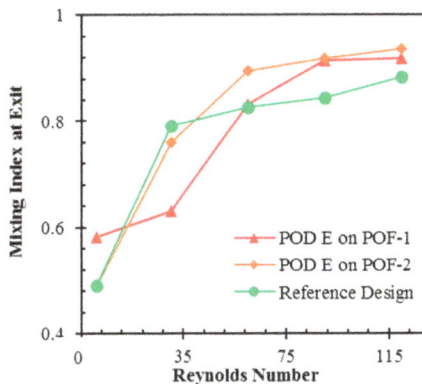

Figure 12. Variations of mixing index at the exit with Reynolds number for POD E on POF-1, POD E on POF-2, and reference design.

In order to compare the merit of PODs at $Re = 1$ with previous micromixers, as shown in Table 7, two different mixing performance parameters named, "mixing cost" [40], and "mixing energy cost" (*mec*) [41] have been used. The terms, "mixing cost" and "mixing energy cost" are defined as:

$$mixing\ cost = \frac{\eta}{\Delta P} \tag{9}$$

$$mec = \frac{\overline{C}_p}{\eta} \tag{10}$$

where η, ΔP, and $\overline{C}_p$ denote mixing efficiency, pressure drop and mean input power coefficient, respectively. The detail description about these terms can be found in the works of Ortega-Casanova [41,42]. Mixing efficiency (η) and mean input power coefficient ($\overline{C}_p$) are defined as follows:

$$\eta = \left(1 - \frac{\sigma}{\sigma_{max}}\right) \times 100 \tag{11}$$

$$\overline{C}_p = 2\Delta Pq/\rho V^2 \tag{12}$$

where σ, σ_{max}, ΔP, q, ρ, and V are the standard deviation of the concentration, maximum standard deviation, pressure drop, dimensionless flow rate, density of fluid, and average flow velocity, respectively. A micromixer with higher mixing cost and lower *mec* indicates a more efficient micromixer. From Table 7, it is evident that PODs have neither the best nor the worst values of these parameters. Higher mixing cost values are shown by all of the PODs as compared to the micromixer proposed by Cheri et al. [43]. PODs show higher *mec* values as compared to the rhombic micromixer with asymmetrical flow [40]. The highest *mec* value is shown by the micromixer configuration of a rectangular chamber with obstruction proposed by Cheri et al. [43]. Although the micromixer proposed by Cheri et al. [43] shows the highest pressure drop, due to the highest mixing efficiency and the lowest mean input power coefficient, the *mec* value becomes least.

Table 7. Comparison of mixing cost and mixing energy cost of PODs on POF-1 with previous micromixers.

						Previous Works			
	PODs on POF-1					Ortega-Casanova [41]	Ortega-Casanova [42]	Chung and Shih [40]	Cheri et al. [43]
Parameters	A	B	C	D	E				
η	44	50	55	59	58	27	27	55	60
ΔP (Pa)	161	199	269	373	391	-*	-*	162	530
$\overline{C}_p$	7273	6800	7141	7619	7624	904	550	10,000	100
mixing cost	0.273	0.251	0.205	0.157	0.149	-*	-*	0.3	0.113
mec	166	137	130	130	131	33	20	182	1.6

-* denotes data not available.

7. Conclusions

Multi-objective optimizations of a micromixer with 3D serpentine and SAR configuration have been performed at Reynolds numbers, 1 and 120, based on flow and mixing analyses using 3D Navier-Stokes equations and convection-diffusion equation. Three design variables, i.e., d/Pi, w_2/Pi, and H/W, were selected, and two objective functions were defined in terms of mixing index at the exit of the micromixer and pressure drop through the micromixer for the optimization. In a parametric study, the mixing index shows maxima for all of the design variables, but the pressure drop shows monotonic variations for all the design variables in the tested ranges. Two concave Pareto-optimal fronts (POF-1 and POF-2) representing trade-off between the two objective functions at $Re = 1$ and 120, respectively, were obtained by RBNN surrogate model and MOGA. In applying RBNN model, it was found that the k-hold cross-validation errors for mixing index and the pressure drop were minimized with the spread constants, 0.9 and 0.5, respectively. On POF-1 ($Re = 1$), the preference of a mixing index-oriented design, POD E over a pressure drop-oriented design, POD A leads to 34.5% relative increase in mixing index at the exit, and the preference of POD A over POD E showed 64.5% reduction in pressure drop. On POF-2 ($Re = 120$), the preference of POD E over POD A leads to only 2.8% relative increase in mixing index at the exit, and the preference of POD A over POD E showed 65.7% reduction in pressure drop. These results indicate that mixing index is less sensitive to the design variables as compared to pressure drop, on the Pareto-optimal fronts. It was found from the numerical analysis at the PODs that the maximum enhancements of 18.8% at POD E on POF-1 and 6.0% at POD E on POF-2 in mixing index at the exit were obtained when compared to the reference design. And, maximum reductions of 5.8% at POD A on POF-1 and 11.1% at POD A on POF-2 in pressure drop were achieved compared to the reference design. Maximum relative error of the surrogate prediction compared to the numerical analysis was smaller for mixing index than pressure drop, and increased with the Reynolds number. In a range of $Re = 1$–120, POD E on POF-1 showed values of mixing index larger than those of the reference design except at $Re = 30$, but POD E on POF-2 showed higher mixing performance than the reference design, only for the Reynolds numbers larger than 60.

Acknowledgments: This work was supported by Inha University research grant.

Author Contributions: Wasim Raza performed numerical analysis, analysis of the results, and drafting the manuscript to be submitted, and Sang-Bum Ma performed numerical optimizations. Kwang-Yong Kim revised the manuscript and gave final authorization of the version to be submitted.

Conflicts of Interest: The authors declare there is no conflict of interest.

Nomenclature

b	length of the vertical sub-channel in O-structure (m)
$\overline{C}_p$	mean input power coefficient
c	mass fraction
D_h	hydraulic diameter of the main channel (m)
d	spacing between two O-structure (m)
F	objective function
H	total depth of the channel (m)
h_1	depth of the horizontal sub-channel in O-structure (m)
h_2	depth of the sub-channel in X-structure (m)
L_o	length of initial part of main channel (m)
L	total length of the mixing unit (m)
L_e	length of the channel outlet (m)
L_t	total length of the micromixer (m)
LHS	Latin hypercube sampling
M	mixing index
M_o	mixing index at the exit
MOGA	multi-objective genetic algorithm
N	number of sampling points
n	number of mixing units
P	pressure (Pa)
ΔP	pressure drop (Pa)
Pi	pitch length (m)
POD	Pareto-optimal design
POF	Pareto-optimal front
p	length of the horizontal sub-channel in O-structure (m)
RBNN	radial basis neural network
Re	Reynolds number
SAR	split and recombine
V	average velocity (m/s)
W	width of the main channel (m)
w_1	width of the sub-channel in O-structure (m)

w_2	width of the sub-channel in X-structure (m)
x, y, z	Cartesian coordinates

Greek Symbols

α	fluid diffusivity coefficient (m^2/s)
η	mixing efficiency
μ	fluid dynamic viscosity (kg·m^{-1}·s^{-1})
ν	fluid Kinematic viscosity (m^2/s)
ρ	fluid density (kg/m^3)
σ	standard deviation

Subscripts

i	sampling point or fluid component
m	optimal mixing
max	maximum value
x	axial distance

References

1. Haeberle, S.; Zengerle, R. Microfluidic platforms for lab-on-a-chip applications. *Lab Chip* **2007**, *7*, 1094–1110. [CrossRef] [PubMed]
2. Pagliara, S.; Dettmer, S.L.; Keyser, U.F. Channel-facilitated diffusion boosted by particle binding at the channel entrance. *Phys. Rev. Lett.* **2014**, *113*. [CrossRef] [PubMed]
3. Lee, C.Y.; Wang, W.T.; Liu, C.C.; Fu, L.M. Passive mixers in microfluidic systems: A review. *Chem. Eng. J.* **2016**, *288*, 146–160. [CrossRef]
4. Gossett, D.R.; Weaver, W.M.; MacH, A.J.; Hur, S.C.; Tse, H.T.K.; Lee, W.; Amini, H.; Di Carlo, D. Label-free cell separation and sorting in microfluidic systems. *Anal. Bioanal. Chem.* **2010**, *397*, 3249–3267. [CrossRef] [PubMed]
5. Nguyen, N.-T.; Wu, Z. Micromixers—A review. *J. Micromech. Microeng.* **2005**, *15*, R1–R16. [CrossRef]
6. Hessel, V.; Löwe, H.; Schönfeld, F. Micromixers—A review on passive and active mixing principles. *Chem. Eng. Sci.* **2005**, *60*, 2479–2501. [CrossRef]
7. Capretto, L.; Cheng, W.; Hill, M.; Zhang, X. Micromixing within microfluidic devices. *Top. Curr. Chem.* **2011**, *304*, 27–68. [PubMed]
8. Liu, A.; He, F.; Wang, K.; Zhou, T.; Lu, Y.; Xia, X. Rapid method for design and fabrication of passive micromixers in microfluidic devices using a direct-printing process. *Lab Chip* **2005**, *5*, 974–978. [CrossRef] [PubMed]
9. Kim, D.S.; Lee, S.H.; Kwon, T.H.; Ahn, C.H. A serpentine laminating micromixer combining splitting/recombination and advection. *Lab Chip* **2005**, *5*, 739–747. [CrossRef] [PubMed]
10. Tofteberg, T.; Skolimowski, M.; Andreassen, E.; Geschke, O. A novel passive micromixer: Lamination in a planar channel system. *Microfluid. Nanofluid.* **2010**, *8*, 209–215. [CrossRef]
11. Buchegger, W.; Wagner, C.; Lendl, B.; Kraft, M.; Vellekoop, M.J. A highly uniform lamination micromixer with wedge shaped inlet channels for time resolved infrared spectroscopy. *Microfluid. Nanofluid.* **2011**, *10*, 889–897. [CrossRef]
12. Nimafar, M.; Viktorov, V.; Martinelli, M. Experimental investigation of split and recombination micromixer in confront with basic T- and O- type micromixers. *Int. J. Mech. Appl.* **2012**, *2*, 61–69. [CrossRef]
13. Chen, X.; Shen, J. Numerical analysis of mixing behaviors of two types of E-shape micromixers. *Int. J. Heat Mass Transf.* **2017**, *106*, 593–600. [CrossRef]
14. Liu, R.H.; Stremler, M.A.; Sharp, K.V.; Olsen, M.G.; Santiago, J.G.; Adrian, R.J.; Aref, H.; Beebe, D.J. Passive mixing in a three-dimensional serpentine microchannel. *J. Microelectromech. Syst.* **2000**, *9*, 190–197. [CrossRef]
15. Stroock, A.D.; Dertinger, S.K.W.; Ajdari, A.; Mezic, I.; Stone, H.A.; Whitesides, G.M. Chaotic mixer for microchannels. *Science* **2002**, *295*, 647–651. [CrossRef] [PubMed]
16. Xia, H.M.; Wan, S.Y.M.; Shu, C.; Chew, Y.T. Chaotic micromixers using two-layer crossing channels to exhibit fast mixing at low Reynolds numbers. *Lab Chip* **2005**, *5*, 748–755. [CrossRef] [PubMed]
17. Feng, X.; Ren, Y.; Jiang, H. An effective splitting-and-recombination micromixer with self-rotated contact surface for wide Reynolds number range applications. *Biomicrofluidics* **2013**, *7*, 1–10. [CrossRef] [PubMed]

18. Le The, H.; Le Thanh, H.; Dong, T.; Ta, B.Q.; Tran-Minh, N.; Karlsen, F. An effective passive micromixer with shifted trapezoidal blades using wide Reynolds number range. *Chem. Eng. Res. Des.* **2015**, *93*, 1–11. [CrossRef]
19. Ansari, M.A.; Kim, K.Y. Application of the radial basis neural network to optimization of a micromixer. *Chem. Eng. Technol.* **2007**, *30*, 962–966. [CrossRef]
20. Ansari, M.A.; Kim, K.Y. Shape optimization of a micromixer with staggered herringbone groove. *Chem. Eng. Sci.* **2007**, *62*, 6687–6695. [CrossRef]
21. Kim, B.S.; Kwak, B.S.; Shin, S.; Lee, S.; Kim, K.M.; Jung, H., II; Cho, H.H. Optimization of microscale vortex generators in a microchannel using advanced response surface method. *Int. J. Heat Mass Transf.* **2011**, *54*, 118–125. [CrossRef]
22. Cortes-Quiroz, C.A.; Zangeneh, M.; Goto, A. On multi-objective optimization of geometry of staggered herringbone micromixer. *Microfluid. Nanofluid.* **2009**, *7*, 29–43. [CrossRef]
23. Afzal, A.; Kim, K.Y. Three-objective optimization of a staggered herringbone micromixer. *Sens. Actuators B Chem.* **2014**, *192*, 350–360. [CrossRef]
24. Hossain, S.; Kim, K.-Y. Optimization of a micromixer with two-layer serpentine crossing channels at multiple Reynolds numbers. *Chem. Eng. Technol.* **2017**, *40*, 1–10. [CrossRef]
25. Raza, W.; Hossain, S.; Kim, K.-Y. Effective mixing in a short serpentine split-and-recombination micromixer. *Sens. Actuators B Chem.* **2018**, *258*, 381–392. [CrossRef]
26. Wang, L.; Yang, J.-T. An overlapping crisscross micromixer using chaotic mixing principles. *J. Micromech. Microeng.* **2006**, *16*, 2684–2691. [CrossRef]
27. ANSYS. *Solver Theory Guide*; CFX-15.0; ANSYS Inc.: Canonsburg, PA, USA, 2013.
28. Bird, R.B.; Stewart, W.E.; Lightfoot, E.N. *Transport Phenomena*; John Wiley & Sons: New York, NY, USA, 1960; ISBN 047107392X.
29. Yang, J.-T.; Huang, K.-J.; Lin, Y.-C. Geometric effects on fluid mixing in passive grooved micromixers. *Lab Chip* **2005**, *5*, 1140. [CrossRef] [PubMed]
30. Chen, X.; Li, T.; Zeng, H.; Hu, Z.; Fu, B. Numerical and experimental investigation on micromixers with serpentine microchannels. *Int. J. Heat Mass Transf.* **2016**, *98*, 131–140. [CrossRef]
31. Kirby, B. *Micro- and Nanoscale Fluid Mechanics: Transport in Microfluidic Devices*; Cambridge University Press: Cambridge, UK, 2010; ISBN 9780521119030.
32. Leonard, B.; Mokhtari, S. *ULTRA-SHARP Nonoscillatory Convection Schemes for High-Speed Steady Multidimensional Flow*; NASA Technical Memorandum; NASA Lewis Research Center: Cleveland, OH, USA, 1990.
33. Liu, M. Computational study of convective-diffusive mixing in a microchannel mixer. *Chem. Eng. Sci.* **2011**, *66*, 2211–2223. [CrossRef]
34. Boss, J. Evaluation of the homegeneity degree of a mixture. *Bulk Solids Handl.* **1986**, *6*, 1207–1215.
35. McKay, M.D.; Beckman, R.J.; Conover, W.J. Comparison of three methods for selecting values of input variables in the analysis of output from a computer code. *Technometrics* **1979**, *21*, 239–245.
36. MATLAB. *The Language of Technical Computing*; Release 14. MathWorks, Inc.: Natick, MA, USA, 2004. Available online: http//www.mathworks.com (accessed on 26 November 2017).
37. Orr, M. Introduction to radial basis function networks. *Univ. Edinbg.* **1996**, 1–7. [CrossRef]
38. Stone, M. Cross-validatory choice and assessment of statistical predictions. *J. R. Stat. Soc.* **1974**, *36*, 111–147.
39. Hossain, S.; Lee, I.; Kim, S.M.; Kim, K.-Y. A micromixer with two-layer serpentine crossing channels having excellent mixing performance at low Reynolds numbers. *Chem. Eng. J.* **2017**, *327*, 268–277. [CrossRef]
40. Chung, C.K.; Shih, T.R. A rhombic micromixer with asymmetrical flow for enhancing mixing. *J. Micromech. Microeng.* **2007**, *17*, 2495. [CrossRef]
41. Ortega-Casanova, J. CFD study on mixing enhancement in a channel at a low reynolds number by pitching a square cylinder. *Comput. Fluids* **2017**, *145*, 141–152. [CrossRef]
42. Ortega-Casanova, J. Enhancing mixing at a very low Reynolds number by a heaving square cylinder. *J. Fluids Struct.* **2016**, *65*, 1–20. [CrossRef]
43. Sadegh Cheri, M.; Latifi, H.; Salehi Moghaddam, M.; Shahraki, H. Simulation and experimental investigation of planar micromixers with short-mixing-length. *Chem. Eng. J.* **2013**, *234*, 247–255. [CrossRef]

micromachines

MDPI

Article

Topology Optimization of Passive Micromixers Based on Lagrangian Mapping Method

Yuchen Guo [1,2], Yifan Xu [3], Yongbo Deng [4] and Zhenyu Liu [1,*]

[1] Changchun Institute of Optics, Fine Mechanics and Physics (CIOMP), Chinese Academy of Science, Changchun 130033, China; guoyuchen15@mails.ucas.edu.cn
[2] University of Chinese Academy of Science, Beijing 100049, China
[3] State Key Laboratory of Mechanical System and Vibration, Shanghai Jiao Tong University, Shanghai 200240, China; xuyifan19890219@sjtu.edu.cn
[4] State Key Laboratory of Applied Optics, Changchun Institute of Optics, Fine Mechanics and Physics (CIOMP), Chinese Academy of Science, Changchun 130033, China; dengyb@ciomp.ac.cn
* Correspondence: liuzy@ciomp.ac.cn; Tel.:+86-431-8670-8138

Received: 26 January 2018; Accepted: 18 March 2018; Published: 20 March 2018

Abstract: This paper presents an optimization-based design method of passive micromixers for immiscible fluids, which means that the Peclet number infinitely large. Based on topology optimization method, an optimization model is constructed to find the optimal layout of the passive micromixers. Being different from the topology optimization methods with Eulerian description of the convection-diffusion dynamics, this proposed method considers the extreme case, where the mixing is dominated completely by the convection with negligible diffusion. In this method, the mixing dynamics is modeled by the mapping method, a Lagrangian description that can deal with the case with convection-dominance. Several numerical examples have been presented to demonstrate the validity of the proposed method.

Keywords: passive micromixer; topology optimization; Lagrangian description; mapping method

1. Introduction

Lab-on-a-chip devices have been widely used in the area of the analysis, synthesis, and separations due to the advantages of high efficiency, portability, and low reagent consumption [1]. Injection, mixing, reaction, cleaning, separation, and detection, which are the functions of conventional analytical laboratory, can be achieved on a centimeter-level chip [2]. Various microfluidic devices have been integrated in lab-on-a-chip, such as micropumps, microvalves, micromixers, and microchannel. Rapid and complete mixing can influence the efficiency of a microfluidic system [3]. Therefore, micromixer plays a significant role in the lab-on-a-chip devices. Based on actuation methods, micromixers can be classified into two categories: active micromixers and passive micromixers [4]. Active mixers use external energy to create chaotic convection, such as pressure [5], magnetohydrodynamics [6], electrokinetics [7] and acoustics [8–10]. Active mixers have short mixing times and distances and can be controlled to be on and off, according to the needs of the users. Because of the requirement of external energy, however, the fabrication and integration of active mixers is complicated and expensive [11]. Comparatively, passive mixers can achieve fluid mixing solely by the geometries of channels and be integrated in a complex microfluidic system simply and directly. The inconsistent cross sections of a microchannel can cause the fluid to be stretched and folded in the transversal direction [12–15]. A reasonable layout of passive micromixer can strengthen the chaotic convection to enhance the mixing performance. Immiscible fluids are widely used in chemical industry, where the particle flow can be considered as the immiscible fluids [16]. The mixture of immiscible fluids appears in biochemical experiments, which is usually not discussed compare to the mixers with phenomenon of convection-diffusion. With the trend of miniaturization in recent years, the micromixers of immiscible fluids with efficient mixing performance are also desired in a microfluidic system.

Topology optimization of fluid flows has been proposed by Borrvall and Petersson for Stokes flow [17] and the Navier-Stokes flow [18–21]. When compared to shape optimization method, the detail topology and shape of the microchannels can be obtained simultaneously by topology optimization. This method has been used to design the microchannel networks, micropumps, no-moving part microvalves, and micromixers [22–25]. In the topology optimization of micromixers, the convection-diffusion equation is usually used to describe the mixing process of the fluids, and objective function is the variance between the actual obtained concentration at the outlet and the expected concentration. All of the topological optimization methods that are mentioned above used to design the mixers of miscible fluids are based on the Eulerian description of the convection-diffusion dynamics and can not be directly applied to the design of immiscible fluid mixers. The mixture of immiscible fluids is a convective problem in physics and should be described by the Navier–Stokes (NS) equations and convection equation in numerical calculation. However, when using the standard Galerkin finite element method to solve the above equation, the numerical instability will occur. For an optimization process, the numerical instability of solving forward problems is a big challenge. To avoid this problem, the mapping method is used in this paper to describe the convection problem in numerical calculation. The mapping method proposed by Singh et al. [26,27] can describe the mixing performance by calculating the mapping matrix and be integrated into the topology optimization method. For the topology optimization problem, the measure of the mixing in the mapping method needs to be discussed. The coarse grained concentration [28] can change the value of the objective function by replacing the concentration in the objective function with the coarse grained concentration defined in discrete areas. In order to shorten the optimization time, the mixing performance of entire periodic structure is obtained by analyzing only one cycle structure in this paper.

In passive mixers, the mixing efficiency is mainly determined by the layout of mixers. The chaotic convection can be promoted by the various cross-sectional structures along the flow direction. Based on the topology optimization method of fluidic flows and Lagrangian description, this paper is focused on the layout design method of the passive micromixers of immiscible fluid, where the mapping method is used to describe the mixing process of the immiscible fluid. A new mixing measurement is applied in this paper, based on the change of position of the mixed fluids. This paper is organized as follows: the choice of mixing measurement and the mapping method are stated in Section 2; the topology optimization model of the passive micromixers and the corresponding adjoint equations and derivative are derived in Section 3; several numerical results are presented in Section 4; and, the discussion and conclusion are stated in Section 5.

2. Measure of Mixing and Lagrangian Mapping Description

2.1. Measure of Mixing

In the topology optimization method of micromixers based on Eulerian description of the convection-diffusion dynamics, we define the concentration as the volume fraction of a fluid in a mixture of two fluids at a point. To quantify and compare mixing performance, the least squares variance between the actual obtained concentration at the outlet and the expected concentration can be expressed as [29]:

$$J(c) = \frac{\int_{\Gamma_{out}} (c - \bar{c})^2 d\Gamma}{\int_{\Gamma_{in}} (c_0 - \bar{c})^2 d\Gamma} \tag{1}$$

where Γ_{in} and Γ_{out} are the inlet and outlet of the mixer, respectively; c_0 is the reference concentration, which is usually the designer specified concentration at the inlet. Ideally, the two fluids are sufficiently mixed in the mixer. Therefore, the expected concentration $\bar{c}$ is chosen as the ideal concentration after sufficient mixing at the outlet, which is the average concentration. The concentrations of two fluids are set to be dimensionless value 0 and 1. When the two fluids are miscible, $J(c) = 0$ means that the two fluids can be considered to be completely mixed and $J(c) = 1$ means that the two fluids can be considered to be completely separated. However, when the two liquids are immiscible, the concentration c of a fluid is always be 0 or 1, due to the absence of diffusion and $J(c)$ remains constant. The expression in Equation (1) is not suitable to quantify the mixing performance. The coarse grain concentration method that was proposed by Welander et al. [28] can avoid this situation. The coarse grain concentration C_i defined on a finite cell Γ_i is:

$$C_i = \int_{\Gamma_i} c(X) d\Gamma \tag{2}$$

where Γ_i is the i-th cell obtained discretely on the cross section perpendicular to the flow direction; C_i can vary in interval [0,1]. Therefore, Equation (1) can be rewrite as

$$J_{\overline{C}}(C) = \frac{1}{A_\Gamma} \sum_{i=1}^{N} \frac{(C_i - \overline{C})^2}{(C_i^0 - \overline{C})^2} A_{\Gamma_i}, \overline{C} = \frac{1}{A_\Gamma} \sum_{i=1}^{N} C_i A_{\Gamma_i} \tag{3}$$

where A_Γ and A_{Γ_i} is the area of cross section and the i-th cell, respectively; $\overline{C}$ is the average coarse grain concentration; C_i^0 is the coarse grain concentration of the i-th cell on the cross section of inlet; and N is the number of cells divided in cross section.

Since the mixed fluids are immiscible, the coarse grain concentration changes only at the contact cells of the two fluids. Due to the low Reynolds numbers of fluid and the limited mixing length of micromixer used in the topology optimization model, the change of the value of mixing performance in Equation (3) is not large enough. When the number of discrete cells increases at the same time, the value of Equation (3) is not sensitive to the change of coarse grain concentration. A new measurement mixing is proposed to amplify the value modification in the mixing performance by using the change of the positions of two fluids at the cross section of outlet. The least squares variance between the actual obtained coarse grain concentration C_i at the outlet and the initial coarse grain concentration C_i^0 at the inlet is used:

$$J_{C^0}(C) = \frac{1}{1 + \frac{1}{A_\Gamma} \sum_{i=1}^{N} (C_i - C_i^0)^2 A_{\Gamma_i}} \tag{4}$$

When $J_{C^0}(C) = 1$, the two immiscible fluids can be considered to be completely separated; when $J_{C^0}(C) < 1$, the two immiscible fluids can be considered to be mixing.

2.2. Mapping Method

A distribution matrix is used in the mapping method to store the information, which describes the changes of fluidic distribution between two specified cross sections [30,31]. To obtain the each coefficient of mapping matrix, the initial cross section is divided into a large number of discrete cells with specified size. The material of fluid transferred to several recipient cells from a donor cell along the fluidic flow. The fraction of material in the recipient cell Ω_j in section at $X = X_0 + \Delta X$, which is found in the donor cell Ω_i in section at $X = X_0$ is the coefficient of mapping matrix. The cross section with N cells can construct a distribution matrix of the order $N \times N$:

$$\varphi_{ij} = \frac{\int \Omega_j|_{x=x_0+\Delta x} \cap \Omega_i|_{x=x_0} dA}{\int \Omega_j|_{x=x_0} dA} \tag{5}$$

To describe the detail of flow, a lot of sections are traced which are set along the flow direction. Tracing all cells in all sections during a flow over a distance ΔX, the detail of flow can be described by the complex deformation of cells. Although this tracking method is feasible, it is too time-consuming to apply into topology optimization. Singh et al. proposed a convenient method for calculating the coefficients of the mapping matrix [26,27]. To approximate the coefficients of mapping matrix,

K markers are filled into each cell uniformly in donor section at $X = X_0$, and then, tracing these markers can obtain the information about the distribution in recipient section at $X = X_0 + \Delta X$. When the number of markers in the cell Ω_j in donor section is M_j and M_{ij} markers are traced in the cell Ω_i in recipient section, the coefficient of mapping matrix can be calculated as

$$\Phi_{ij} = \frac{M_{ij}}{M_j} \tag{6}$$

The convection dynamics of fluids can be analyzed the following procedure. Using the same cell to describe the coarse grain concentration, the coarse grain concentration distribution of cross-section can construct a vector $\mathbf{C} \in R^{N \times 1}$ (N is the number of cells). After passing through the structure that is described by the mapping method, the coarse grain concentration distribution of recipient section $\mathbf{C}^r$ can be calculated from the coarse grain concentration distribution of donor section $\mathbf{C}^d$ as:

$$\mathbf{C}^r = \Phi \mathbf{C}^d \tag{7}$$

The coarse grain concentration distribution at outlet can be obtained as:

$$\mathbf{C}_o = \Phi_{all} \mathbf{C}_i \tag{8}$$

where $\mathbf{C}_i$ and $\mathbf{C}_o$ are the coarse grain concentration distribution of inlet and outlet cross-section, respectively; and Φ_{all} is the mapping matrix of whole mixer.

Since the construction of the mapping matrix has no correlation with the initial cross-section concentration distribution, the mixer of periodic layout, with a known initial mapping matrix Φ^1 of a single cycle, has the following relation:

$$\mathbf{C}^{i+1} = \Phi^1 \mathbf{C}^i, \mathbf{C}^n = \underbrace{(\Phi^1(\Phi^1(\ldots(\Phi^1 \mathbf{C}^0))))}_{n \text{ times}} \tag{9}$$

where $\mathbf{C}^i$ is the coarse grain concentration distribution after the i-th mixing. Therefore, for the periodic mixers in any cycle, the computational cost is saved that the concentration distribution. $\mathbf{C}^n$ can be obtained by simply multiplying the single cycle matrix in corresponding times.

Assuming that the length direction of mixing channel is the x-axis, the cross section that is shown in Figure 1 is the y-z section. The trajectories of markers are tracked based on the coordinate axis. Tracing can be realized by the axial velocity component u_x and the transversal velocity components u_y and u_z, respectively:

$$\frac{dy}{dx} = \frac{u_y}{u_x}, \frac{dz}{dx} = \frac{u_z}{u_x} \tag{10}$$

Since the particles become disordered at any downstream position in forward tracing method, there is no guarantee that an equal number of traced markers can be found in each cell at the outlet. Using the backward particle tracing (BTP) to construct the distribution matrix, the markers initially fill the recipient cell in the cross section of outlet, and they are traced backward against the flow direction [26,27],

$$X_i = X_0 + \sum_{i=1}^{n} (\int_{x_i+\Delta x}^{x_i} \frac{\mathbf{u}}{\mathbf{u}^T \mathbf{I_u}} dX) \tag{11}$$

where $X_i = (x_i, y_i, z_i)$ and $X_0 = (x_0, y_0, z_0)$ are the coordinates of the markers at the cross section of inlet and outlet, respectively; $\mathbf{I_u} = [1\ 0\ 0]^T$; $n = n_{all} - 1$, n_{all} is the number of cross section in the mapping method. Due to the integration of axial spatial increment rather than time, the error that is caused by the different time distribution can be eliminated. However, this approach is not effective when the fluid is reflowing.

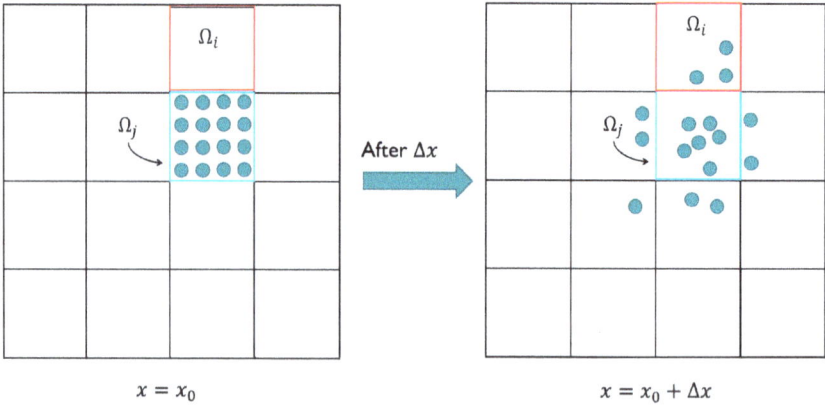

Figure 1. The calculation of the mapping coefficient Φ_{ij} in the mapping matrix is the ratio of the number of markers received by the recipient cell Ω_i at $X = X_0 + \Delta X$ to the initial number of markers in Ω_j at $X = X_0$ (in this example $\Phi_{ij} = 3\,/16$) [26,27].

Figure 2 shows a top view of the grooves on the bottom in a staggered herringbone mixer (SHM). We apply the geometry of SHM and the material properties used in the study of Singh et al. [26]. The length ratio of two arms of every groove is 2:1, and all arms are at 45° to the axial direction; the channel height h = 77 μm, the channel width w = 200 μm, the depth of grooves g_d = 17.7 μm, the width of grooves g_w = 70.7 μm and the distance between two grooves also equals 70.7 μm; the viscosity and density of the fluid in the micromixers are 0.067 kg·m/s and 1.2×10^3 kg/m³. The average inlet velocity u = 0.2 cm/s. The velocity field is obtained by 144,000 hexahedral elements and 155,031 nodes. Figure 3a–c show the mixing evolutions in a SHM consisting of one groove type, whose mapping matrix is represented by Φ^1. In the mapping computations of Figure 3a,b, the tracing cross section is covered with 50 × 50 and 100 × 100 cells, respectively, and each cell is filled with 100 uniformly distributed markers. Figure 3c shows the results in the study of Singh et al. [26] using the above parameters. Figure 3d shows the coarse grain concentration distribution at outlet of SHM with ten grooves. The results in Figure 3 demonstrated the validity of the mapping method that is used in this paper. These parameters of mapping method will be used in the optimization process.

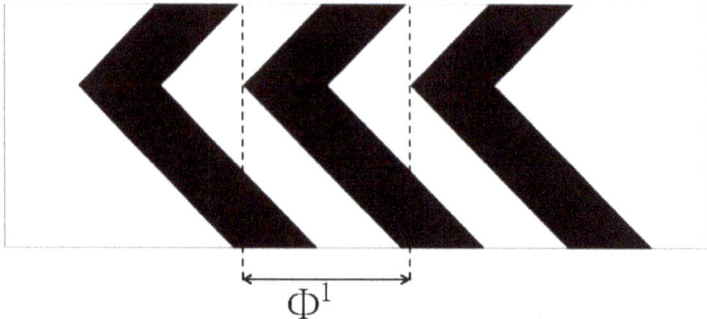

Figure 2. Schematic representation of the grooves in a staggered herringbone mixer (SHM). The mapping matrix Φ^1 covers a single groove applying a fully developed velocity field.

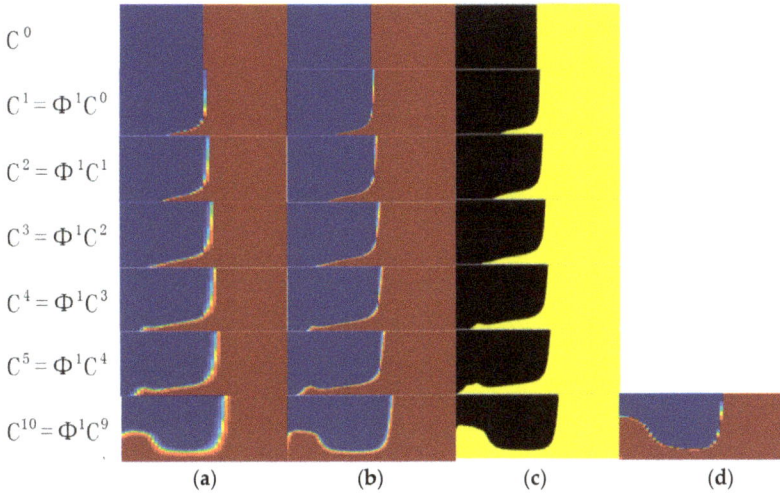

Figure 3. The evolution of coarse grain concentration distribution C^i in a SHM with one groove type: (**a**) The result of mapping method which the tracing cross section is covered with 50×50 cells, and each cell is filled with 100 uniformly distributed markers; (**b**) The result of mapping method which the tracing cross section is covered with 100×100 cells, and each cell is filled with 100 uniformly distributed markers; (**c**) The results in the study of Singh et al. which the tracing cross section is covered with 200×200 cells, and each cell is filled with 256 uniformly distributed markers [26]; (**d**) The coarse grain concentration distribution at outlet of SHM with ten grooves.

3. Topology Optimization Model of Mixers

When the area of each cell divided by the mapping method is the same, the Equation (4) can be simplified to

$$J_{C^0}(C) = \frac{N}{N + \sum\limits_{i=1}^{N} (C_i - C_i^0)^2} \tag{12}$$

where N is the number of cells divided in cross section. This discrete equation is used as quantitative criterion to measure the mixing performance of the immiscible fluids. In this paper, the key point is how to find a reasonable layout of a micromixer in a designer specified design domain, which minimizes the quantitative criterion and mixing can be obtained using the topology optimization method. Based on the continuity assumption, the fluidic field will be described by the Navier-Stokes equations:

$$\rho(\mathbf{u}\cdot\nabla)\mathbf{u} - \eta\Delta\mathbf{u} + \nabla p = \mathbf{f}$$
$$-\nabla\cdot\mathbf{u} = 0 \tag{13}$$

where u, p, ρ, and η are the velocity, pressure, density, and viscosity of the fluid, respectively; f is the body forces acting on the fluid. In the topology optimization method, an artificial friction force is introduced into the Navier-Stokes equations, which was proposed for the Stokes flow by Borrvall and Petersson [17] and generalized to the Navier-Stokes flow by Gersborg et al. [20] and Olesen et al. [21]. Initially, an artificial porous material is uniformly distributed in the design domain; and then, the artificial porous material forms solid and liquid phases gradually; at last, the high and low impermeability characterize the solid phase and the fluid phase, respectively. The artificial friction force is $f = -\alpha(\gamma)u$, and α is the impermeability of the artificial porous material, and γ is the design variable. The design variable γ varies in interval [0,1], where 0 and 1 denote the solid and fluid phases, respectively. The impermeability of porous material α is the interpolation function of design variable γ [17–22]:

$$\alpha(\gamma) = \alpha_{min} + (\alpha_{max} - \alpha_{min})\frac{q(1-\gamma)}{(q+\gamma)} \tag{14}$$

where α_{max} and α_{min} are the impermeability of the solid phase and the fluid phase, respectively; q is a positive value used to adjust the convexity of the interpolation function; α_{min} is chosen 0 in the fluidic topology optimization. To obtain the perfect impermeability of solid no-slip boundary, α_{max} should be infinite, but a finite number has to be chosen to ensure the numerical stability. The layout of the passive micromixer of immiscible fluid can be determined by seeking the distribution of the design variable γ. The topology optimization model based on mapping method is:

$$\min : J_{C0}(C) = \frac{N}{N + \sum\limits_{i=1}^{N}(C_i - C_i^0)^2}$$

$$s.t. \rho(\mathbf{u}\cdot\nabla)\mathbf{u} - \eta\Delta\mathbf{u} + \nabla p = -\alpha\mathbf{u} \quad \text{in } \Omega$$

$$-\nabla\cdot\mathbf{u} = 0 \quad \text{in } \Omega$$

$$\mathbf{u} = \mathbf{u}_0 \quad \text{on } \Gamma_{in}$$

$$\left[-p\mathbf{I} + \eta(\nabla\mathbf{u} + (\nabla\mathbf{u})^{\mathsf{T}})\right]\cdot\mathbf{n} = 0 \quad \text{on } \Gamma_{out}$$

$$0 \leq \gamma \leq 1 \tag{15}$$

$$X = X_0 + \sum\limits_{i=0}^{n}(\int_{x_i + \Delta x}^{x_i}\frac{\mathbf{u}}{\mathbf{u}^{\mathsf{T}}\mathbf{I}_\mathbf{u}}dX) \quad \text{on } \Gamma_{in}$$

$$X = X_0 \quad \text{on } \Gamma_{out}$$

$$C = C_0 \quad \text{on } \Gamma_{in}$$

$$C = \Phi(y_i, z_i)C_0 \quad \text{on } \Gamma_{out}$$

$$\Phi_{ij} = \frac{M_{ij}}{M_j}$$

where $\Omega = \Omega_D \cup \Omega_C$ is the computational domain; Ω_D is the design domain; Ω_C is the channels connected to the inlet, outlet, and design domain.

The constraint optimization model in Equation (15) can be transferred into unconstrained one by Lagrangian multiplier method:

$$L = \quad J + a(\mathbf{u}, \lambda_\mathbf{u})_\Omega + \rho((\mathbf{u}\cdot\nabla)\mathbf{u}, \lambda_\mathbf{u})_\Omega + (\nabla p, \lambda_\mathbf{u})_\Omega + (\alpha\mathbf{u}, \lambda_\mathbf{u})_\Omega$$

$$-(\mathbf{g}, \lambda_\mathbf{u})_{\Gamma_N} + (\mathbf{u} - \mathbf{u}_0, \lambda_\mathbf{u})_{\Gamma_D} + (-\nabla\cdot\mathbf{u}, \lambda_p)_\Omega + (C - \Phi(y, z)C_0, \lambda_C)_{\Gamma_N} \tag{16}$$

$$+(C - C_0, \lambda_C)_{\Gamma_D} + (X - X_0, \lambda_X)_{\Gamma_D} + (X_i - X_0 - \sum\limits_{i=0}^{n}(\int_{x_i + \Delta x}^{x_i}\frac{\mathbf{u}}{\mathbf{u}^{\mathsf{T}}\mathbf{I}_\mathbf{u}}dX), \lambda_X)_{\Gamma_N}$$

where $a(\mathbf{u}, \lambda_\mathbf{u})_\Omega = \int_\Omega \eta\nabla\mathbf{u} : \nabla\lambda_\mathbf{u}d\Omega$; $(*, *)_\Omega$ and $(*, *)_\Gamma$ is the inner product on the computational domain and boundary; $\lambda_\mathbf{u}, \lambda_p, \lambda_C$ and λ_X are the adjoint variable of the velocity field, the pressure field, the coarse grain density on the cross outlet section and the coordinate after particle tracing, respectively.

The variation of the Equation (16) is

$$\delta L = \frac{\partial L}{\partial \mathbf{u}}\delta\mathbf{u} + \frac{\partial L}{\partial p}\delta p + \frac{\partial L}{\partial X}\delta X + \frac{\partial L}{\partial C}\delta C + \frac{\partial L}{\partial \gamma}\delta\gamma \tag{17}$$

According to the Karush-Kuhn-Tucker conditions for the partial differential equation constrained optimization problem, the optimization problem in Equation (15) can be obtained by solving the following adjoint equations [19] (see the Appendix A for the detailed derivation):

$$
\begin{cases}
-\eta\Delta\lambda_{\mathbf{u}} - \rho(\mathbf{u}\cdot\nabla)\lambda_{\mathbf{u}} + \rho(\nabla\mathbf{u})\cdot\lambda_{\mathbf{u}} + \nabla\lambda_p = -\alpha\lambda_{\mathbf{u}} \quad \text{in } \Omega \\
-\nabla\cdot\lambda_{\mathbf{u}} = 0 \quad \text{in } \Omega \\
\lambda_{\mathbf{u}} = 0 \quad \text{on } \Gamma_D \\
[\eta(\nabla\lambda_{\mathbf{u}} + (\nabla\lambda_{\mathbf{u}})^{\mathsf{T}}) - \lambda_p\mathbf{I}]\cdot\mathbf{n} = -\rho(\mathbf{u}\cdot\mathbf{n})\lambda_{\mathbf{u}} + \sum_{i=0}^{n}(\int_{x_0+\Delta x}^{x_0}(\frac{1}{\mathbf{u}^{\mathsf{T}}\mathbf{I}_{\mathbf{u}}}\mathbf{I} - \frac{\mathbf{u}}{(\mathbf{u}^{\mathsf{T}}\mathbf{I}_{\mathbf{u}})^2}\mathbf{I}_{\mathbf{u}}^{\mathsf{T}})dX)\lambda_X \quad \text{on } \Gamma_N \\
\lambda_X = \frac{\partial\Phi}{\partial X_i}\lambda_C \quad \text{on } \Gamma_N \\
\lambda_X = 0 \quad \text{on } \Gamma_D \\
\lambda_C = 0 \quad \text{on } \Gamma_D \\
\lambda_C = -\dfrac{-2N(C - C^0)}{[N + \sum\limits_{i=1}^{N}(C_i - C_i^0)^2]^2} \quad \text{on } \Gamma_N
\end{cases}
\tag{18}
$$

The adjoint sensitivity of the optimization model in Equation (15) is:

$$
\frac{DL}{D\gamma}\Big|_\Omega = \frac{\partial\alpha}{\partial\gamma}\mathbf{u}\cdot\lambda_{\mathbf{u}}
\tag{19}
$$

When considering the manufacturability, the micromixer is designed to be a single layer structure, like the SHM [12–14,32,33]. In the optimization model, an additional constraint is added to ensure that the design variable in the depth direction have a consistent value. The design variable is defined on the plane $x0y$ (Figure 4), and the adjoint derivative of the optimization model (19) is changed to [22]:

$$
\frac{DL}{D\gamma} = \frac{1}{z_t - z_b}\int_{z_b}^{z_t}\frac{\partial\alpha}{\partial\gamma}\mathbf{u}\cdot\lambda_{\mathbf{u}}dz
\tag{20}
$$

where z_t and z_b are the coordinates in z direction of top and bottom surface of the design domain, and $z_t > z_b$.

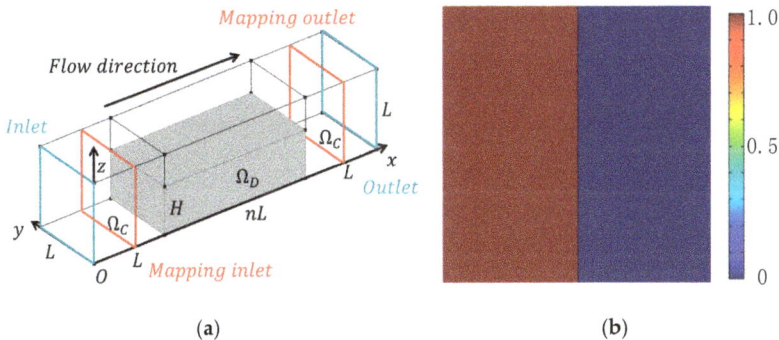

Figure 4. (a) Design domain Ω_D of the micromixers at the bottom layer of the straight channel. Ω_C is the channels connected to the inlet, outlet and design domain Ω_D. The length of design domain Ω_D is n times of L; (b) the distribution of coarse grain concentration on the outlet.

The topology optimization problem is solved by using the gradient-based iterative approach. In the optimization iterations, the Navier-Stokes equations, the backward particles tracing equation, and the corresponding adjoint equations in the weak form are solved by the finite element method using the commercial software COMSOL Multiphysics (version 3.5, COMSOL Inc., Stockholm, Sweden). The method of moving asymptotes (MMA) is used to update the design variable. The optimization iterations are stopped, when the maximal change of the objective value in three consecutive iterations less than 1×10^{-3}. The procedure of solving the topology optimization problem for the layout design of passive micromixers is listed below.

1. Give the initial value of the design variable γ;
2. Solve the Navier-Stokes equations and backward particle tracing equation by the finite element method;
3. Solve the adjoint equation;
4. Compute the adjoint derivative and the corresponding objective and constraint values;
5. Update the design variable by method of moving asymptotes (MMA);
6. Check for convergence; if the stopping conditions are not satisfied, go to 2; and
7. Post-processing

4. Results and Discussion

To demonstrate the validity of the proposed method, passive micromixers of immiscible fluid in a straight microchannel with external driven flow (constant flow rate at inlet) is investigated in the following. The design domain is shown in Figure 4a, and each cubic space with an edge size equals to L is discretized by $20 \times 20 \times 20$ hexahedral elements. The width of micromixer L is 400 µm and the height of design domain H is 240 µm. The viscosity and density of the fluid in the micromixers are 1×10^{-3} Pa·s and 1×10^3 kg/m³. The value of α_{max} and q in the topology method are chosen 1×10^7 and 0.1. The initial value of the design variable γ is 0.4, which should be between [0,1], as shown in Equation (15). Since the backward particle tracing method is applied, the number of cell in the cross section of outlet is 50×50 and 100 markers are traced in each cell. The coarse grain concentrations of two immiscible fluids are set to be dimensionless value 1 and 0 (Figure 4b), respectively. All of the computations are performed on a DELL workstation (DELL Optiplex 7040, Intel Core i7-6700 CPU, 16 gigabyte memory, Round Rock, TX, USA).

Based on the topology optimization model in Section 3, the optimal layouts of the passive micromixers of immiscible fluid with different Reynolds number and length of the design domain are obtained, as shown in Figures 5–7. The optimized results strongly depend on the parameters, such as the size of design domain and the Re number. From the comparison between the results in Figures 5 and 6, the obtained layouts of micromixers depend on the selection of Reynolds number. Different values of n are chosen to compare the mixing performance of the micromixers obtained by the topology optimization method (Figures 6 and 7). A larger value of *n* means a longer mixing length and a more reasonable layout in the micromixers. When the length of the mixer is relatively short, the herringbone type structure is not necessary; when the length of single-periodic becomes longer, the herringbone type structure promotes the mixing effect obviously. By comparing the optimal results, the herringbone type structure is a reasonable topology, which could be used further for detailed shape or size optimization when the mixer has an appropriate length.

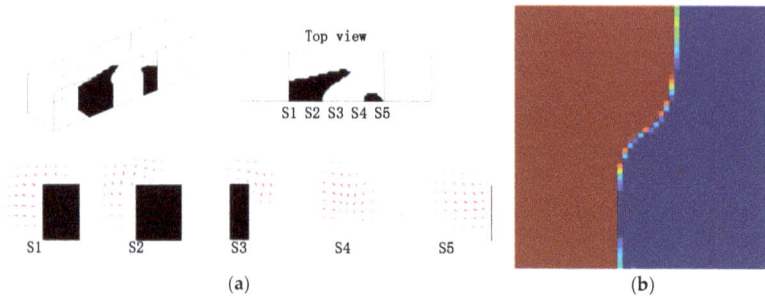

Figure 5. (**a**) Layout of the micromixer for the design domain as shown in Figure 4, where *n* = 2, *Re* = 2.5, and projected velocity vector distribution in the cross sections (S1, S2, S3, S4, S5); the mixing measurement $J_{C_0}(C)$ = 0.9141; (**b**) the distribution of coarse grain concentration on the outlet.

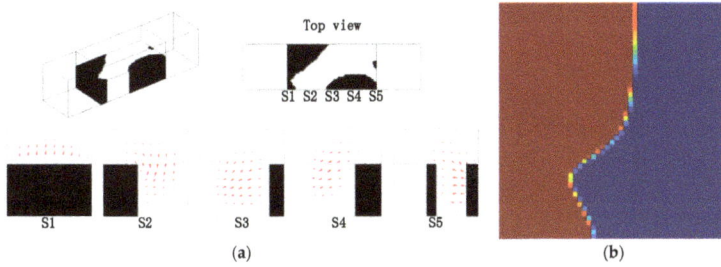

Figure 6. (**a**) Layout of the micromixer for the design domain as shown in Figure 4, where n = 2, Re = 5, and projected velocity vector distribution in the cross sections (S1, S2, S3, S4, S5); the mixing measurement $J_{C_0}(C)$ = 0.9147; (**b**) the distribution of coarse grain concentration on the outlet.

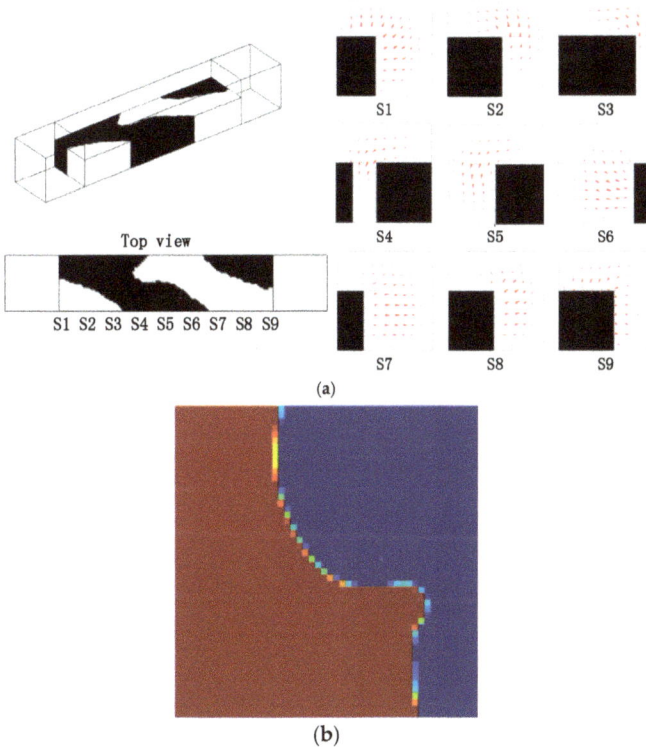

Figure 7. (**a**) Layout of the micromixer for the design domain as shown in Figure 4, where n = 4, Re = 5, and projected velocity vector distribution in the cross sections (S1, S2, S3, S4, S5, S6, S7, S8, S9); the mixing measurement $J_{C_0}(C)$ = 0.8382; (**b**) the distribution of coarse grain concentration on the outlet.

Mapping method provides an approximate way to obtain the mixing performance of mixer with spatial periodic layout by multiplying the single cycle mapping matrix in corresponding times. Figure 8 shows the evolution of mixing performance of micromixers with different cycles where the single cycle is the layout in Figure 5. Due to the Equation (4) is only valid for situations where fluid agitation is not obvious, we used Equation (3) to measure the mixing performance. One can see that the mixing effect is enhanced as the number of cycle increases. The effect of numerical diffusion becomes apparent, when the multiplying times of mapping matrix increase. However, a mutual comparison is still possible and valuable.

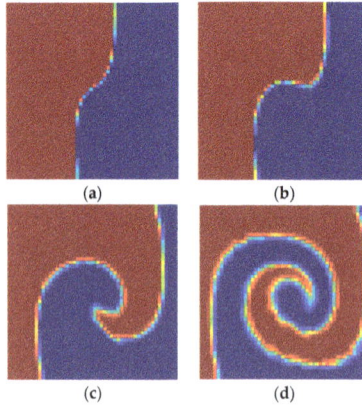

Figure 8. The evolution of mixing performance of micromixers with different cycles which the single cycle is the layout in Figure 5: (**a–d**) are the distribution of coarse grain concentration on the outlet with 1, 2, 5, and 10 cycles, and $J_{\overline{C}}(C)$ are 0.9882, 0.9779, 0.9404 and 0.8413, respectively.

The CPU time in one optimization iteration step is 500 s for the case that design domain is shown in Figure 4, with $n = 2$ and $Re = 5$. Therefore, designing a multi-period structure directly means having a longer computational domain and more CPU time. In contrast, the mapping method can obtain multi-period mixing performance easily by simply multiplying the single cycle matrix in corresponding times (Figure 8). Figure 9 shows the mixing performance of the micromixer that is shown in Figure 5 and SHM shown in Figure 10, which has only one groove and same volume as the micromixer shown in Figure 5. Therefore, there are full of room to adjust the expression of objective in the optimization model (Equation (15)) proposed in this paper. Whereas, the derivation of adjoint sensitivity is unchanged in the case that the chosen objective function is differentiable for the design variable. The whole optimization procedure is valid for immiscible mixer design, as illustrated in this paper.

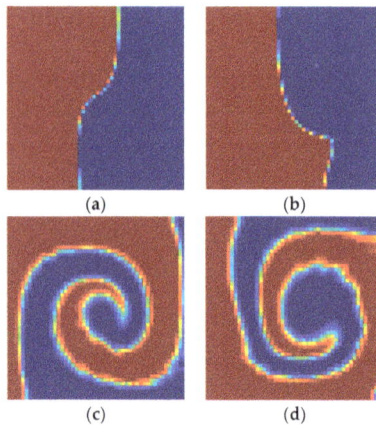

Figure 9. (**a,c**) are the distribution of coarse grain concentration on the outlet with 1 and 10 cycles shown in Figure 8a and d, and $J_{\overline{C}}(C)$ are 0.9882 and 0.8413; (**b,d**) are the distribution of coarse grain concentration on the outlet of SHM (Figure 10) with 1 and 10 cycles, and $J_{\overline{C}}(C)$ are 0.9883 and 0.8446.

Micromachines **2018**, *9*, 137

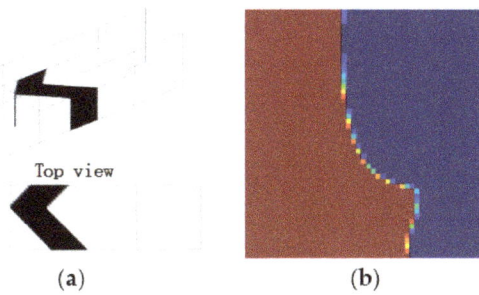

Figure 10. (a) Layout of the SHM in the design domain as shown in Figure 4, where $n = 2$, $Re = 2.5$, and the mixing measurement $J_{\overline{C}}(C) = 0.9883$; (b) the distribution of coarse grain concentration on the outlet. (the SHM has only one groove and is same volume as the micromixer shown in Figure 5).

5. Conclusions

In this paper, a novel method is used to design the layout of the passive micromixers of immiscible fluid has been proposed based on the topology optimization of fluidic flows. The layout of the passive micromixers is determined by solving a topology optimization problem to minimize the mixing measurement. Additionally, the detailed mixing performance is obtained by using mapping method in Lagrangian description without consideration of the diffusion. The mapping method used in this paper is a rough approximated method. However, it can produce results that are in good agreement with the conventional numerical simulation method of up to 10 cycles. The discrepancies may increase with the using of more cycles. The design variable is set to represent the impermeability distribution of the artificial porous medium in the design domain. Based on the adjoint analysis of the topology optimization problem, the design variable is evolved to derive impermeability distribution of the artificial porous medium with low and high levels, which, respectively, correspond to the solid and fluid phases in the micromixer. The manufacturability of the obtained layout is ensured by the manufacturing constraint. Numerical results demonstrated the validity of the proposed method. In addition, this method can be extended to design micromixers with multi-layer structure and active micromixers. These will be investigated in our future work.

Acknowledgments: This research was funded by the National Science Foundation of China under grant Nos. 51675506, 51405465. The authors are grateful to Krister Svanberg for supply of the Matlab codes of MMA.

Author Contributions: Yifan Xu, Yuchen Guo, Yongbo Deng and Zhenyu Liu wrote the topology optimization code; Yongbo Deng, Yifan Xu, Zhenyu Liu and Yuchen Guo derived the theoretical formulas; Yuchen Guo and Zhenyu Liu performed the numerical examples; Yuchen Guo, Zhenyu Liu and Yongbo Deng wrote the paper.

Conflicts of Interest: The authors declare no conflict of interest.

Appendix A

Transfer the constraint model in Equation (15) into unconstrained format by Lagrangian multiplier method:

$$
\begin{aligned}
L = \ & J + a(\mathbf{u}, \lambda_{\mathbf{u}})_\Omega + \rho((\mathbf{u}\cdot\nabla)\mathbf{u}, \lambda_{\mathbf{u}})_\Omega + (\nabla p, \lambda_{\mathbf{u}})_\Omega + (\alpha\mathbf{u}, \lambda_{\mathbf{u}})_\Omega \\
& - (\mathbf{g}, \lambda_{\mathbf{u}})_{\Gamma_N} + (\mathbf{u} - \mathbf{u}_0, \lambda_{\mathbf{u}})_{\Gamma_D} + (-\nabla\cdot\mathbf{u}, \lambda_p)_\Omega + (C - \Phi(y,z)C_0, \lambda_C)_{\Gamma_N} \\
& + (C - C_0, \lambda_C)_{\Gamma_D} + (X - X_0, \lambda_X)_{\Gamma_D} + \Big(X_i - X_0 - \sum_{i=0}^{n} \big(\int_{x_i+\Delta x}^{x_i} \tfrac{\mathbf{u}}{\mathbf{u}^T \mathbf{I}_{\mathbf{u}}} dX\big), \lambda_X\Big)_{\Gamma_N}
\end{aligned}
\tag{A1}
$$

where $a(\mathbf{u}, \lambda_{\mathbf{u}})_\Omega = \int_\Omega \eta\nabla\mathbf{u} : \nabla\lambda_{\mathbf{u}} d\Omega$; $(*,*)_\Omega$ and $(*,*)_\Gamma$ is the inner product on the computational domain and boundary; $\lambda_{\mathbf{u}}$, λ_p, λ_C, and λ_X are the adjoint variables of the velocity field, the pressure field, the coarse grain density on the cross outlet section and the coordinate after particle tracing, respectively. λ_u and λ_C are vectors as well as λ_p is scalar. $\lambda_X = (\lambda_x \ \lambda_y \ \lambda_z)^T$. The variation of the main function Equation (A1) is:

$$\delta L = \frac{\partial L}{\partial \mathbf{u}}\delta \mathbf{u} + \frac{\partial L}{\partial p}\delta p + \frac{\partial L}{\partial X_i}\delta X_i + \frac{\partial L}{\partial C}\delta C + \frac{\partial L}{\partial \gamma}\delta \gamma \tag{A2}$$

According to the Karush-Kuhn-Tucker conditions for the partial differential equation constrained optimization problem, the optimization model in Equation (15) can be solved by solving the following adjoint equations:

$$\frac{\partial L}{\partial \mathbf{u}}\delta \mathbf{u} = 0 \tag{A3}$$

$$\frac{\partial L}{\partial p}\delta p = 0 \tag{A4}$$

$$\frac{\partial L}{\partial X_i}\delta X_i = 0 \tag{A5}$$

$$\frac{\partial L}{\partial X_i}\delta X_i = 0 \tag{A6}$$

For the Equation (A3):

$$\frac{\partial L}{\partial \mathbf{u}}\delta \mathbf{u} = a(\eta \delta \mathbf{u}, \lambda_{\mathbf{u}})_{\Omega} + \rho \delta ((\mathbf{u} \cdot \nabla)\mathbf{u}, \lambda_{\mathbf{u}})_{\Omega} + (\alpha \delta \mathbf{u}, \lambda_{\mathbf{u}})_{\Omega} - (\nabla \cdot \delta \mathbf{u}, \lambda_p)_{\Omega}$$
$$+ (\delta \mathbf{u}, \lambda_{\mathbf{u}})_{\Gamma_D} - \left(\int_{x_0+\Delta x}^{x_0} \left(\frac{1}{\mathbf{u}^T \mathbf{I}_{\mathbf{u}}}\mathbf{I} - \frac{\mathbf{u}}{(\mathbf{u}^T \mathbf{I}_{\mathbf{u}})^2}\mathbf{I}_{\mathbf{u}}^T \right) dx, \lambda_X \right)_{\Gamma_N} \tag{A7}$$

where:

$$a(\eta \delta \mathbf{u}, \lambda_{\mathbf{u}})_{\Omega} = -(\eta \Delta \lambda_{\mathbf{u}}, \delta \mathbf{u})_{\Omega} + (\eta (\nabla \lambda_{\mathbf{u}} + (\nabla \lambda_{\mathbf{u}})^T) \cdot \mathbf{n}, \delta \mathbf{u})_{\Gamma_N} \tag{A8}$$

$$\delta ((\mathbf{u} \cdot \nabla)\mathbf{u}, \lambda_{\mathbf{u}})_{\Omega} = -((\mathbf{u} \cdot \nabla)\lambda_{\mathbf{u}}, \delta \mathbf{u})_{\Omega} + (\nabla \mathbf{u} \cdot \lambda_{\mathbf{u}}, \delta \mathbf{u})_{\Omega} + ((\mathbf{u} \cdot \mathbf{n})\lambda_{\mathbf{u}}, \delta \mathbf{u})_{\Gamma_N} \tag{A9}$$

$$(\nabla \cdot \delta \mathbf{u}, \lambda_p)_{\Omega} = -(\nabla \lambda_p, \delta \mathbf{u})_{\Omega} + (\lambda_p \mathbf{n}, \delta \mathbf{u})_{\Gamma_N} \tag{A10}$$

By inserting Equations (A8)–(A10) into Equation (A7), one can obtain:

$$\frac{\partial L}{\partial \mathbf{u}}\delta \mathbf{u} = \int_{\Omega} [-\eta \Delta \lambda_{\mathbf{u}} - \rho (\mathbf{u} \cdot \nabla)\lambda_{\mathbf{u}} + \rho (\nabla \mathbf{u}) \cdot \lambda_{\mathbf{u}} + \nabla \lambda_p + \alpha \lambda_{\mathbf{u}}] \cdot \delta \mathbf{u} d\Omega + \int_{\Gamma_D} [\lambda_{\mathbf{u}}] \cdot \delta \mathbf{u} d\Gamma$$
$$+ \int_{\Gamma_N} [(\eta (\nabla \lambda_{\mathbf{u}} + (\nabla \lambda_{\mathbf{u}})^T - \lambda_p \mathbf{I}) \cdot \mathbf{n} + \rho (\mathbf{u} \cdot \mathbf{n})\lambda_{\mathbf{u}} - \sum_{i=0}^{n} (\int_{x_0+\Delta x}^{x_0} (\frac{1}{\mathbf{u}^T \mathbf{I}_{\mathbf{u}}}\mathbf{I} - \frac{\mathbf{u}}{(\mathbf{u}^T \mathbf{I}_{\mathbf{u}})^2}\mathbf{I}_{\mathbf{u}}^T) dX)\lambda_X] \cdot \delta \mathbf{u} d\Gamma \tag{A11}$$
$$= 0$$

For the Equation (A4):

$$\frac{\partial L}{\partial p}\delta p = \int_{\Omega} [-\nabla \cdot \lambda_{\mathbf{u}}] \cdot \delta p d\Omega = 0 \tag{A12}$$

For the Equation (A5):

$$\frac{\partial L}{\partial X_i}\delta X_i = \int_{\Gamma_N} [\lambda_X - \frac{\partial \Phi}{\partial X_i}\lambda_C] \cdot \delta X_i d\Gamma + \int_{\Gamma_D} [\lambda_X] \cdot \delta X_i d\Gamma = 0 \tag{A13}$$

For the Equation (A6):

$$\frac{\partial L}{\partial C}\delta C = \int_{\Gamma_N} \left[\frac{-2N(C-C^0)}{[N + \sum\limits_{i=1}^{N}(C_i - C_i^0)^2]^2} + \lambda_C \right] \cdot \delta C d\Gamma + \int_{\Gamma_D} [\lambda_C] \cdot \delta C d\Gamma = 0 \tag{A14}$$

Finally, the adjoint equation can be obtained as:

$$
\begin{cases}
-\eta\Delta\lambda_{\mathbf{u}} - \rho(\mathbf{u}\cdot\nabla)\lambda_{\mathbf{u}} + \rho(\nabla\mathbf{u})\cdot\lambda_{\mathbf{u}} + \nabla\lambda_p = -\alpha\lambda_{\mathbf{u}} \quad \text{in } \Omega \\
-\nabla\cdot\lambda_{\mathbf{u}} = 0 \quad \text{in } \Omega \\
\lambda_{\mathbf{u}} = 0 \quad \text{on } \Gamma_D \\
[\eta(\nabla\lambda_{\mathbf{u}} + (\nabla\lambda_{\mathbf{u}})^{\mathsf{T}}) - \lambda_p\mathbf{I}]\cdot\mathbf{n} = -\rho(\mathbf{u}\cdot\mathbf{n})\lambda_{\mathbf{u}} + \sum_{i=0}^{n} (\int_{x_0+\Delta x}^{x_0} (\frac{1}{\mathbf{u}^{\mathsf{T}}\mathbf{I_u}}\mathbf{I} - \frac{\mathbf{u}}{(\mathbf{u}^{\mathsf{T}}\mathbf{I_u})^2}\mathbf{I_u^T})dX)\lambda_X \quad \text{on } \Gamma_N \\
\lambda_X = \frac{\partial\Phi}{\partial X_i}\lambda_C \quad \text{on } \Gamma_N \\
\lambda_X = 0 \quad \text{on } \Gamma_D \\
\lambda_C = 0 \quad \text{on } \Gamma_D \\
\lambda_c = -\frac{-2N(C-C^0)}{[N+\sum_{i=1}^{N}(C_i-C_i^0)^2]^2} \quad \text{on } \Gamma_N
\end{cases}
\tag{A15}
$$

According to the Equations (A3)–(A6) into the variation Equation (A2) of main function, one can obtain:

$$
\delta L = \frac{\partial L}{\partial\gamma}\delta\gamma = \int_{\Omega} [\frac{\partial\alpha}{\partial\gamma}\mathbf{u}\cdot\lambda_{\mathbf{u}}]\delta\gamma d\Omega
\tag{A16}
$$

Then the adjoint derivative of the optimization model in Equation (15) is:

$$
\frac{DL}{D\gamma}\Big|_{\Omega} = \frac{\partial\alpha}{\partial\gamma}\mathbf{u}\cdot\lambda_{\mathbf{u}}
\tag{A17}
$$

References

1. Whitesides, G. The lab finally comes to the chip! *Lab Chip* **2014**, *14*, 3125–3126. [CrossRef] [PubMed]
2. Manz, A.; Graber, N.; Widmer, H.M. Miniaturized total chemical-analysis systems—A novel concept for chemical sensing. *Sens. Actuator B-Chem.* **1990**, *1*, 244–248. [CrossRef]
3. Lee, C.-Y.; Chang, C.-L.; Wang, Y.-N.; Fu, L.-M. Microfluidic mixing: A review. *Int. J. Mol. Sci.* **2011**, *12*, 3263–3287. [CrossRef] [PubMed]
4. Suh, Y.K.; Kang, S. A review on mixing in microfluidics. *Micromachines* **2010**, *1*, 82–111. [CrossRef]
5. Glasgow, I.; Aubry, N. Enhancement of microfluidic mixing using time pulsing. *Lab Chip* **2003**, *3*, 114–120. [CrossRef] [PubMed]
6. Qian, S.; Zhu, J.; Bau, H.H. A stirrer for magnetohydrodynamically controlled minute fluidic networks. *Phys. Fluids* **2002**, *14*, 3584–3592. [CrossRef]
7. Chen, L.; Deng, Y.; Zhou, T.; Pan, H.; Liu, Z. A novel elecrtoosmotic micromixer with asymmetric lateral structures and DC electrode arrays. *Micromachines* **2017**, *8*, 105. [CrossRef]
8. Ozcelik, A.; Ahmed, D.; Xie, Y.; Nama, N.; Qu, Z.; Nawaz, A.A.; Huang, T.J. An acoustofluidic micromixer via bubble inception and cavitation from microchannel sidewalls. *Anal. Chem.* **2014**, *86*, 5083–5088. [CrossRef] [PubMed]
9. Huang, P.H.; Xie, Y.; Ahmed, D.; Rufo, J.; Nama, N.; Chen, Y.; Chan, C.Y.; Huang, T.J. An acoustofluidic micromixer based on oscillating sidewall sharp-edges. *Lab Chip* **2013**, *13*, 3847–3852. [CrossRef] [PubMed]
10. Nama, N.; Huang, P.H.; Huang, T.J.; Costanzo, F. Investigation of acoustic streaming patterns around oscillating sharp edges. *Lab Chip* **2014**, *14*, 2824–2836. [CrossRef] [PubMed]
11. Nguyen, N.-T.; Wu, Z. Micromixers—A review. *J. Micromech. Microeng.* **2005**, *15*, R1–R16. [CrossRef]
12. Stroock, A.D.; Dertinger, S.W.; Ajdari, A.; Mezić, I.; Stone, A.; Whitesides, G.M. Chaotic mixer for microchannels. *Science* **2002**, *295*, 647–651. [CrossRef] [PubMed]
13. Liu, Y.; Kim, B.J.; Sung, H.J. Two-fluid mixing in a microchannel. *Int. J. Heat Fluid Flow* **2004**, *25*, 986–995. [CrossRef]
14. Camesasca, M.; Kaufman, M.; Manaszloczower, I. Staggered passive micromixers with fractal surface patterning. *J. Micromech. Microeng.* **2006**, *16*, 2298–2311. [CrossRef]
15. Lee, S.W.; Kim, D.S.; Lee, S.S.; Kwon, T.H. A split and recombination micromixer fabricated in a PDMS three-dimensional structure. *J. Micromech. Microeng.* **2006**, *16*, 1067–1072. [CrossRef]

16. Meijer, H.E.H.; Singh, M.K.; Anderson, P.D. On the performance of static mixers: A quantitative comparison. *Prog. Polym. Sci.* **2012**, *37*, 1333–1349. [CrossRef]
17. Borrvall, T.; Petersson, J. Topology optimization of fluids in stokes flow. *Int. J. Numer. Methods Fluids* **2003**, *41*, 77–107. [CrossRef]
18. Deng, Y.; Liu, Z.; Wu, Y. Topology optimization of steady and unsteady incompressible Navier-Stokes flows driven by body forces. *Struct. Multidiscip. Optim.* **2013**, *47*, 555–570. [CrossRef]
19. Deng, Y.; Liu, Z.; Zhang, P.; Liu, Y.; Wu, Y. Topology optimization of unsteady incompressible Navier-Stokes flows. *J. Comput. Phys.* **2011**, *230*, 6688–6708. [CrossRef]
20. Gersborg-Hansen, A.; Sigmund, O.; Haber, R.B. Topology optimization of channel flow problems. *Struct. Multidiscip. Optim.* **2005**, *30*, 181–192. [CrossRef]
21. Olesen, L.H.; Okkels, F.; Bruus, H. A high-level programming language implantation of topology optimization applied to steady-state Navier-Stokes flow. *Int. J. Numer. Methods Eng.* **2006**, *65*, 975–1001. [CrossRef]
22. Deng, Y.; Liu, Z.; Zhang, P.; Liu, Y.; Gao, Q.; Wu, Y. A flexible layout design method for passive micromixers. *Biomed. Microdevices* **2012**, *14*, 929–945. [CrossRef] [PubMed]
23. Zhou, T.; Wang, H.; Shi, L.; Liu, Z.; Joo, S. An enhanced eletroosmotic micromixer with an efficient asymmetric lateral structure. *Micromachines* **2016**, *7*, 218. [CrossRef]
24. Ji, Y.; Deng, Y.; Liu, Z.; Zhou, T.; Wu, Y.; Qian, S. Optimal control-based inverse determination of electrode distribution for electroosmotic micromixer. *Micromachines* **2017**, *8*, 247. [CrossRef]
25. Andreasen, C.; Gersborg, A.; Sigmund, O. Topology optimization of microfluidic mixers. *Int. J. Numer. Meth. Fluids* **2009**, *61*, 498–513. [CrossRef]
26. Singh, M.K.; Kang, T.G.; Meijer, H.E.H.; Anderson, P.D. The mapping method as a toolbox to analyze, design, and optimize micromixers. *Microfluid. Nanofluid* **2008**, *5*, 313–325. [CrossRef]
27. Singh, M.K.; Galaktionov, O.S.; Meijer, H.E.H.; Anderson, P.D. A simplified approach to compute distribution matrices for the mapping method. *Comput. Chem. Eng.* **2009**, *33*, 1354–1362. [CrossRef]
28. Welander, P. Studies on the general development of motion in a two-dimensional ideal fluid. *Tellus* **1955**, *7*, 141–156. [CrossRef]
29. Danckwerts, P.V. The definition and measurement of some characteristics of mixtures. *Appl. Sci. Res.* **1952**, *3*, 279–296. [CrossRef]
30. Kruijt, P.G.M.; Galaktionov, O.S.; Anderson, P.D.; Peters, G.W.M.; Meijer, H.E.H. Analyzing mixing in periodic flows by distribution matrices: Mapping method. *AIChE J.* **2001**, *47*, 1005–1015. [CrossRef]
31. Kruijt, P.G.M.; Galaktionov, O.S.; Peters, G.W.M.; Meijer, H.E.H. The mapping method for mixing optimizatiom. Part II Transport in a Corotating Twin Screw Extruder. *Int. Polym. Proc.* **2001**, *16*, 161–171. [CrossRef]
32. Du, Y.; Zhang, Z.; Yim, C.H.; Lin, M.; Cao, X. A simplified design of the staggered herringbone micromixer. *Biomicrofluidics* **2010**, *4*, 024105. [CrossRef] [PubMed]
33. Williams, K.J.; Longmuir, K.J.; Yager, P. A practical guide to the staggered herringbone mixer. *Lab Chip* **2008**, *8*, 1121–1129. [CrossRef] [PubMed]

MDPI

Article

3D Multi-Microchannel Helical Mixer Fabricated by Femtosecond Laser inside Fused Silica

Chao Shan [1,2], Feng Chen [1,2,*], Qing Yang [2,3], Zhuangde Jiang [2,3] and Xun Hou [1]

[1]   State Key Laboratory for Manufacturing System Engineering and Key Laboratory of Photonics Technology for Information of Shanxi Province, School of Electronics & Information Engineering, Xi'an Jiaotong University, Xi'an 710049, China; shanchaobest@stu.xjtu.edu.cn (C.S.); houxun@mail.xjtu.edu.cn (X.H.)

[2]   The International Joint Research Center for Micro/Nano Manufacturing and Measurement Technologies, Xi'an Jiaotong University, Xi'an 710049, China; yangqing@mail.xjtu.edu.cn (Q.Y.); zdjiang@mail.xjtu.edu.cn (Z.J.)

[3]   School of Mechanical Engineering, Xi'an Jiaotong University, Xi'an 710049, China

*   Correspondence: chenfeng@mail.xjtu.edu.cn; Tel.: +86-029-8266-8420

Received: 22 November 2017; Accepted: 12 January 2018; Published: 16 January 2018

Abstract: Three-dimensional (3D) multi-microchannel mixers can meet the requirements of different combinations according to actual needs. Rapid and simple creation of 3D multi-microchannel mixers in a "lab-on-a-chip" platform is a significant challenge in micromachining. In order to realize the complex mixing functions of microfluidic chips, we fabricated two kinds of complex structure micromixers for multiple substance mixes simultaneously, separately, and in proper order. The 3D multi-microchannel mixers are fabricated by femtosecond laser wet etch technology inside fused silica. The 3D multi-microchannel helical mixers have desirable uniformity and consistency, which will greatly expand their utility and scope of application.

Keywords: passive micromixer; femtosecond laser; microstructure fabrication

---

1. Introduction

Micro-total analysis systems and "lab-on-a-chip" platforms are widely used for sample preparation and analysis, drug delivery, and biological and chemical synthesis [1–5]. As essential components of many microfluidic systems, micro-mixing devices are used to homogenize samples or reactions. With the development of microfluidic systems, numerous micromixers based on kind of material have been reported [6–10].

At present, the application of microfluidic chips allows micromixers to mix only two substances at once, hardly meeting new requirements. Therefore, new micromixers that can mix multiple substances simultaneously or separately are needed. Sometimes, mixing substances in the proper order is also required. In chemistry and biology, there is a lot of demand for these more complex micro-mixers. Especially in the area of biological cell culture, cells are often mixed with a variety of nutrient solutions [11]. There are also greater requirements for the micromixer function. Complex structure design can realize complex functions. Therefore, it is necessary to fabricate a multi-microchannel mixer to meet the requirements of different combinations.

Microfluidic devices are not only prepared in fused silica, but also are processed on polymer materials [12,13]. Polymer materials have poor stability and resistance to acid and alkali. Fused silica is an ideal material for microfluidic applications; it has very high optical transparency, low background fluorescence, chemical inertness, and hydrophilicity [14]. Therefore, preparing microfluidic devices inside fused silica can be very effective. However, it is still challenging to fabricate three-dimensional (3D) microfluidic chips in hard/brittle materials like fused silica. Research on micromixer fabrication is crucial to microfluidic chip development.

Micromixers are commonly fabricated by a conventional photolithography and chemical etching process. Planar lithography techniques can be used to prepare some simple micromixers. The reason why this is insufficient for fabricating multichannel micromixers is that the steps of multilayer stacked structure methods become more complicated when the processing structures are asymmetric or complex. Additionally, this approach lacks flexibility. The planar or semi-three-dimensional structure of micromixers limits the mix performance of the devices [15,16]. As a consequence, rapid and efficient mixing is still a challenging task in the design and development of micromixers.

Based on the demand, femtosecond laser micromachining has become a feasible tool to fabricate 3D structures inside transparent hard/brittle materials [17,18]. In a previous study [19], we proposed fabrication of 3D microchannels with arbitrary lengths and uniform diameters in fused silica by femtosecond laser wet etch (FLWE) technology. This simple and flexible technology provides new ideas and solutions for processing 3D multi-microchannel mixers in fused silica.

Compared with previous work, the difficulty of multichannel micromixer processing mainly exists in device uniformity, integration level, and stability. The channel diameters of single micromixers fabricated in previous work have inhomogeneity of several microns. The defect of the encapsulation process also makes the mix process fail often. In order to address these deficiencies, we took a variety of approaches: adopted more accurate dynamic laser power regulation and two-step wet etch method for device uniformity; adopted the method of oxygen plasma assisted encapsulation for device stability; and designed two different structures for different mix requirements. We also paid more attention to practical applications of 3D helical mixers and the device properties and functionality.

With these approaches, we fabricated 3D multi-microchannel helical mixers with good uniformity and consistency inside fused silica based on FLWE technology. The design and improved micromixer structure can extend the performance of the chip to meet diverse needs. This work addresses the two main problems with creating feasible 3D multi-microchannel mixers: (1) effectively improving the device's integrality; and (2) developing a robust but simple packaging method to connect the 3D micromixers to the external devices. In addition, the utility and scope of application of the micromixers fabricated by FLWE are greatly improved.

2. Materials and Methods

The micromixer manufacturing process includes two steps: the laser writing process and the chemical wet etch, as shown in Figure 1. The femtosecond laser micromachining system used to fabricate the micromixer involves a femtosecond laser source (wavelength: 800 nm, pulse duration: 50 fs, repetition rate: 1 kHz), a microscope objective (NA = 0.9, 100×, Nikon, Tokyo, Japan), a programmable three-axis stage (H101A ProScan II Upright Stage, Prior Scientific, Cambridge, United Kingdom), a charge-coupled device camera (Nikon), and a laser beam control system. The helical micromixer was fabricated embedded in the fused-silica substrate (1.0 cm × 1.0 cm × 0.9 mm). The channel was written by the laser by moving the 3D stage along the pattern path at a speed of 10 μm/s. The laser power was adjusted from 3 mW to 7 mW by a computer-controlled attenuator with temporal modulation of the power compensation. Instead of linearly polarized laser light, circularly polarized laser light is used in the manufacture of complex 3D microchannels with a high etching rate [20]. The geometry of the channel, the height, and the pitch circle diameter of the helix can be controlled by computer.

Second, we used two-step wet etch. The mixer sample was immersed in an ultrasound-assisted solution of 5% hydrofluoric acid (HF) for 1 h. We call this process pre-etching. This allows the entire femtosecond laser–irradiated pattern to have sufficient, but not intense, contact with HF. Then we put this sample into another solution of 10% HF for about 2 h, and a hollow spiral channel with a dimension of 40 μm was obtained. With this approach, it is possible to improve the uniformity and smoothness of the microchannel's inner wall to reduce the flow resistance of liquid materials.

To create the connector for the injection process, a fused-silica sample was placed on a prepared polydimethylsiloxane (PDMS) film. We optimally treated the PDMS substrate and the silica chip with oxygen plasma for 80 s, and then bonded them together under room temperature conditions. Finally, two syringe needles were inserted into the PDMS film at the entrance of the channel to connect all the channels. Thus, all the channels will penetrate as integrated.

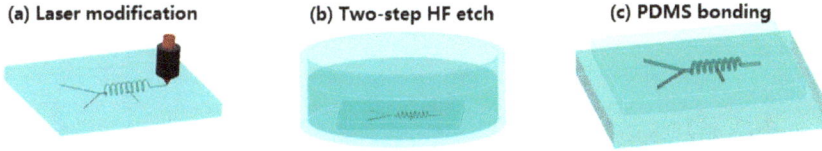

Figure 1. Schematic diagram of the fabrication process. (**a**) Laser modification with accurate dynamic laser power regulation; (**b**) 5% HF wet etch for 1 h and 10% HF wet etch for 2 h; and (**c**) oxygen plasma assisted encapsulation.

3. Results

In order to realize the complex mixing function for multiple substance mixes simultaneously, separately, or in proper order, we designed and prepared two complex structure micromixers.

Parallel Mixer

The first structure, shown in Figure 2, is composed of multiple channels to mix substances simultaneously or separately. The structure of the micromixer group is made up of two helical microchannel structures (A, B), three inlets (a, b, c), two outlets (d, e), and many microchannels. For the helical microchannel, the number of turns, length, helical pitch, circle diameter, and depth of the helix axis with respect to the sample surface are 12, 1200, 100, 150, and 175 μm, respectively. The diameter of the cross-section is 35 ± 1 μm. In previous studies, there was inhomogeneity of several microns in size. However, in this work, the size error of a single helical structure can be controlled to be between 1 and 2 μm, which greatly improves the uniformity of the device. Meanwhile, the size error of multiple helical structures can be controlled to be between 1 and 2 μm, which ensures consistency of the device. This improvement is achieved by adopting more efficient power and accurately adjusting the scanning speed. The power and scanning speed of the machining process are adjusted according to the structure.

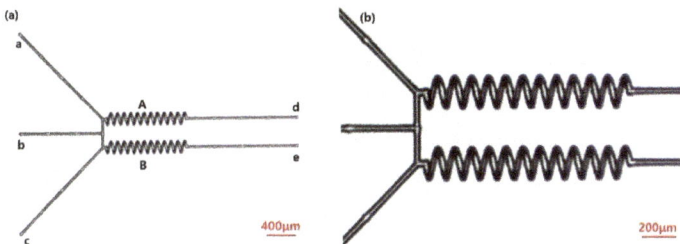

Figure 2. (**a**) Morphology of a multiple channel mixer, observed via optical microscope; and (**b**) detail of mixer with enlarged scale.

The micromixer can complete the mixing of substances (material 1 and material 2) through the two inlets (a and b) in helical microchannel (A). It can complete the mixing of substances (material 2 and material 3) through the two inlets (b and c) in helical microchannel (B). It can also complete the mixing of substances (material 1 and material 3) in helical channel A or B. More importantly, the micromixer can mix three substances simultaneously.

Traditional mixers are usually combined in series. The mixing of various samples is generally implemented through multiple mixers, step by step. This increases the number of microfluidic chip devices and decreases integration. However, with the multichannel micromixer group we prepared, the utilization efficiency of the micromixer can be greatly improved, and the flexibility of the application is also greatly enhanced. This micromixer can complete different mix tasks at the same time, reducing the number of micromixers needed and further simplifying the structure of microfluidic chips.

For the multiple mixers group, as well, integration is an important performance index. Femtosecond laser micro-nano processing technology can improve integration of the microfluidic chip. On the one hand, in previous studies we were able to obtain significantly higher mixing efficiency of the 3D helical mixer than of the straight channel, and also reduce the size of the chip to save space [19]. On the other hand, reducing the spacing between each unit is an important way to improve integration. As can be seen in Figure 1, the spacing between the two spiral channels is only 100 µm.

In order to further show the ability of femtosecond laser micro-nano processing technology to improve the degree of integration, we designed another parallel micromixer like the above structure. We can control the distance between two spiral microchannels to be only a few microns. Greatly reducing the chip size improves integration of the chip. As shown in Figure 3, the space is <5 µm. More importantly, we can integrate multiple micromixers with other devices, such as a micro cabin, or even a metal microstructure [21], to realize more complex functions.

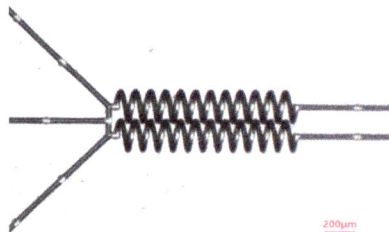

Figure 3. Small spacing structure of a multiple channel mixer.

The second micromixer structure we designed is shown in Figure 4. It is a micromixer used to mix different substances in a particular order. It consists of three inlets (a, b, c), a spiral microchannel, and one outlet (d). For the helical microchannel, the number of turns, length, helical pitch, circle diameter, and depth of the helix axis with respect to the sample surface are 12, 1200, 100, 150, and 175 µm, respectively. The diameter of the cross-section is 35 ± 1 µm.

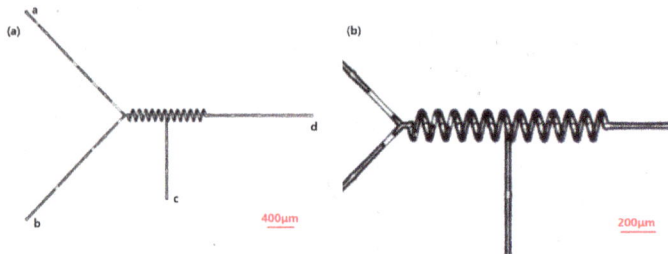

Figure 4. (**a**) Morphology of a multiple channel mixer, observed via an optical microscope; and (**b**) detail of a mixer with enlarged scale.

This mixer can first mix material 1 and material 2 in a spiral microchannel through two inlets, a and b. After mixing evenly, it can continue to mix or react with material 3 in the second half of the spiral microchannel. This step-by-step process has special applications in certain chemical and biological fields. This is very important in the expansion of mix functions. At the same time, we can add new entrances to other parts of the spiral microchannel to meet the multistep mixing function.

4. Discussion

We conducted a simple test of the mixing performance of the functional mixer for specific effects of functional mixing. As shown in Figure 5a, NaOH solution (concentration 0.1 M) and phenolphthalein solution (concentration 0.05 M) were injected simultaneously into the two entrances, a and b, respectively, by a syringe pump at a speed of 0.1 mL/h (phenolphthalein becomes pink once mixed with a basic NaOH solution), corresponding to a flow rate of 8.3 cm/s and a low Reynolds number, Re ≈ 2.5. The c inlet did not have any substance injected. As we can see, in the spiral microchannel, the color changes to red, which proves that the two solutions are mixed. As shown in Figure 5b, after the previous step, we injected diluted hydrochloric acid (concentration 0.1 M) into the c inlet. It can be seen in the figure that in the first half of the spiral channel it remained red, while in the second half it became colorless. This proves that sodium hydroxide solution and diluted hydrochloric acid are neutralized and the phenolphthalein becomes colorless, thus proving the mixing effect of the liquid in the second half of the spiral channel.

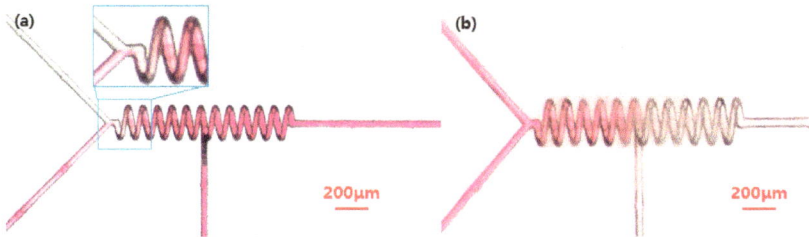

Figure 5. (**a**) Optical micrograph of mixing results of NaOH solution and phenolphthalein solution; (**b**) Mixing results after injecting diluted hydrochloric acid.

Finally, the color intensity was obtained from the captured images, pixel by pixel. The intensity of each pixel was normalized. The normalized values correspond to the degree of mixing for each pixel, 0 for unmixed and 1 for fully mixed. The mixing index σ was calculated as:

$$I_{ni} = 1 - \frac{I_i - I_{mix}}{I_{unmix} - I_{mix}} \tag{1}$$

$$\sigma = 1 - \sqrt{\frac{1}{N} \sum_{n=1}^{N} (1 - I_{ni})^2} \tag{2}$$

where I_{ni} is the normalized pixel intensity, I_i is the pixel intensity, I_{unmix} is the intensity at the unmixed region, I_{mix} is the intensity at the fully mixed region, and N is the number of pixels.

Mixing of the solution is very slow in the upriver straight microchannel. A sharp increase of mixing occurs at the joint of the helical microchannel and the straight microchannel, where a chaotic flow is supposed to arise, as shown in Figure 5a. From the mixing index σ of the normalized pixel intensity I_{ni}, as illustrated in Figure 6. Here, we define the intersection of the Y-channel as the x-coordinate origin. The straight channel is about 100 μm. Then, full mixing ($\sigma \geq 0.9$) can be verified in 400 μm (at the beginning of the third turn) along the micromixer. The experiment shows that full mixing can be achieved in the helical micromixer quickly and with high efficiency.

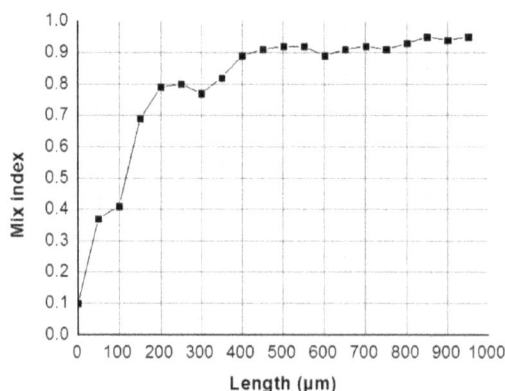

Figure 6. Experimental mixing performance of the helical micromixer: mixing index σ graphics of Figure 5a.

5. Conclusions

In summary, by means of FLWE technology, some complex 3D multi-microchannel mixers can be achieved inside fused silica. Thus, multi-microchannel mixers with high integration and uniformity for high-performance applications are realized, demonstrating the flexibility and universality of FLWE technology. Meanwhile, a broad spectrum of microfluidics systems fabricated based on these compact and complex 3D multi-microchannel mixers shows promise.

Acknowledgments: This work is supported by the National Key Research and Development Program of China under grant No. 2017YFB1104700, the National Science Foundation of China under grant Nos. 51335008 and 61475124, the NSAF grant No. U1630111, the Collaborative Innovation Center of Suzhou Nano Science and Technology. The SEM work was done at International Center for Dielectric Research (ICDR), Xi'an Jiaotong University.

Author Contributions: Chao Shan and Feng Chen conceived and designed the experiments; Chao Shan performed the experiment and manuscript writing; Qing Yang and Zhuangde Jiang contributed reagents and materials; Feng Chen and Xun Hou performed document revising.

Conflicts of Interest: The authors declare no conflict of interest.

References

1. Chin, C.D.; Laksanasopin, T.; Cheung, Y.K. Microfluidics-based diagnostics of infectious diseases in the developing world. *Nat. Med.* **2011**, *17*, 1015–1019. [CrossRef] [PubMed]
2. Jeong, G.S.; Chung, S.; Kim, C.B. Applications of micromixing technology. *Analyst* **2010**, *135*, 460–473. [CrossRef] [PubMed]
3. Blazej, R.G.; Kumaresan, P.; Mathies, R.A. Microfabricated bioprocessor for integrated nanoliter-scale sanger DNA sequencing. *Proc. Natl. Acad. Sci. USA* **2006**, *103*, 7240–7245. [CrossRef] [PubMed]
4. Whitesides, G.M. The origins and the future of microfluidics. *Nature* **2006**, *442*, 368–373. [CrossRef] [PubMed]
5. Demello, A.J. Control and detection of chemical reactions in microfluidic systems. *Nature* **2006**, *442*, 394–402. [CrossRef] [PubMed]
6. Fang, W.F.; Yang, J.T. A novel microreactor with 3D rotating flow to boost fluid reaction and mixing of viscous fluids. *Sens. Actuators B-Chem.* **2009**, *140*, 629–642. [CrossRef]
7. Sadabadi, H.; Packirisamy, M.; Wuthrich, R. High performance cascaded PDMS micromixer based on split-and-recombination flows for lab-on-a-chip applications. *RSC Adv.* **2013**, *3*, 7296–7305. [CrossRef]

8.   Lee, S.W.; Lee, S.S. Rotation effect in split and recombination micromixing. *Sens. Actuators B-Chem.* **2008**, *129*, 364–371. [CrossRef]

9.   Nimafar, M.; Viktorov, V.; Martinelli, M. Experimental investigation of split and recombination micromixer in confront with basic T- and O- type micromixers. *Int. J. Mech. Appl.* **2012**, *2*, 61–69. [CrossRef]

10. Lee, K.; Kim, C.; Shin, K.S. Fabrication of round channels using the surface tension of PDMS and its application to a 3D serpentine mixer. *J. Micromech. Microeng.* **2007**, *17*, 1533–1541. [CrossRef]

11. Lee, N.Y.; Yamada, M.; Seki, M. Development of a passive micromixer based on repeated fluid twisting and flattening, and its application to DNA purification. *Anal. Bioanal. Chem.* **2005**, *383*, 776–782. [CrossRef] [PubMed]

12. Verma, M.K.S.; Ganneboyina, S.R.; Vinayak Rakshith, R.; Ghatak, A. Three-dimensional multi helical microfluidic mixers for rapid mixing of liquids. *Langmuir* **2008**, *24*, 2248–2251. [CrossRef] [PubMed]

13. Jani, J.M.; Wessling, M.; Lammertink, R.G.H. Geometrical influence on mixing in helical porous membrane microcontactors. *J. Membr. Sci.* **2011**, *378*, 351–358. [CrossRef]

14. Osellame, R.; Hoekstra, H.J.W.M.; Cerullo, G. Femtosecond laser microstructuring: An enabling tool for optofluidic lab-on-chips. *Laser Photonics Rev.* **2011**, *5*, 442–463. [CrossRef]

15. Chung, C.K.; Shih, T.R.; Chen, T.C. Mixing behavior of the rhombic micromixers over a wide Reynolds number range using Taguchi method and 3D numerical simulations. *Biomed. Microdevices* **2008**, *10*, 739–748. [CrossRef] [PubMed]

16. Xia, H.M.; Shu, C.; Wan, S.Y.M. Influence of the Reynolds number on chaotic mixing in a spatially periodic micromixer and its characterization using dynamical system techniques. *J. Micromech. Microeng.* **2006**, *16*, 53–61. [CrossRef]

17. Sugioka, K.; Xu, J.; Wu, D. Femtosecond laser 3D micromachining: A powerful tool for the fabrication of microfluidic, optofluidic, and electrofluidic devices based on glass. *Lab Chip* **2014**, *14*, 3447–3458. [CrossRef] [PubMed]

18. Chen, F.; Shan, C.; Liu, K. Process for the fabrication of complex three-dimensional microcoils in fused silica. *Opt. Lett.* **2013**, *38*, 2911–2914. [CrossRef] [PubMed]

19. He, S.; Chen, F.; Liu, K.; Yang, Q. Fabrication of three-dimensional helical microchannels with arbitrary length and uniform diameter inside fused silica. *Opt. Lett.* **2012**, *37*, 3825–3827. [CrossRef] [PubMed]

20. Yu, X.; Liao, Y.; He, F.; Zeng, B.; Cheng, Y. Tuning etch selectivity of fused silica irradiated by femtosecond laser pulses by controlling polarization of the writing pulses. *J. Appl. Phys.* **2011**, *109*, 053114. [CrossRef]

21. Bian, H.; Shan, C.; Chen, F. Miniaturized 3-D solenoid-type micro-heaters in coordination with 3-D microfluidics. *J. Microelectromech. Syst.* **2017**, *26*, 588–592. [CrossRef]

micromachines

MDPI

Review

A Review on Micromixers

Gaozhe Cai [1], Li Xue [1], Huilin Zhang [1] and Jianhan Lin [2,*]

[1] Key Laboratory of Agricultural Information Acquisition Technology (Beijing) of Ministry of Agriculture, China Agricultural University, 17 East Qinghua Road, Beijing 100083, China; gaozhe@cau.edu.cn (G.C.); li_xue@cau.edu.cn (L.X.); huilinzhang@cau.edu.cn (H.Z.)

[2] Modern Precision Agriculture System Integration Research Key Laboratory of Ministry of Education, China Agricultural University, 17 East Qinghua Road, Beijing 100083, China

* Correspondence: jianhan@cau.edu.cn; Tel.: +86-10-6273-7599

Received: 7 August 2017; Accepted: 1 September 2017; Published: 9 September 2017

Abstract: Microfluidic devices have attracted increasing attention in the fields of biomedical diagnostics, food safety control, environmental protection, and animal epidemic prevention. Micromixing has a considerable impact on the efficiency and sensitivity of microfluidic devices. This work reviews recent advances on the passive and active micromixers for the development of various microfluidic chips. Recently reported active micromixers driven by pressure fields, electrical fields, sound fields, magnetic fields, and thermal fields, etc. and passive micromixers, which owned two-dimensional obstacles, unbalanced collisions, spiral and convergence-divergence structures or three-dimensional lamination and spiral structures, were summarized and discussed. The future trends for micromixers to combine with 3D printing and paper channel were brought forth as well.

Keywords: passive micromixer; active micromixer; microfluidic mixing; microfluidic device

1. Introduction

In the past decade, various microfluidic or lab-on-a-chip [1] devices have been attempted the analysis of biological and chemical targets in the fields of biomedical diagnostics, food safety control, environmental protection, and animal epidemic prevention, etc., and have received increasing attention due to their compact size, automatic operation, faster detection, less reagent, higher sensitivity and in-field use. They can generally integrate injection, mixing, reaction, washing, separation and detection onto a centimeter-level chip [2]. Micromixers, which have a considerable impact on the efficiency and sensitivity of microfluidic devices, are one of the most important components of these devices. Unlike the macro-scale fluidic devices where the mixing of fluids often relies on convection effects, mixing in the micro-scale fluidic ones is often achieved in the microchannels with external turbulences and/or special microstructures at micro-level dimensions to obtain larger surface-to-volume ratio and increasing heat and mass transfer efficiency. Besides, the flow rates of the fluids are generally very low in the microfluidic devices and the regime of the fluids in the microchannels are basically laminar flow with the Reynolds number of <1, indicating the fluid flows in parallel layers with no disruption between the layers and the mixing of the fluids is mainly dependent on diffusion with a very low mixing efficiency. For example, in a water-based (a fluid density of 1 kg/m^3 and a viscosity of 0.001 $N \cdot s/m^2$) microfluidic system with a channel width of 100 μm and a flow rate of 1 μL/s, the Reynolds number is 0.1 and it takes 1 s for the fluids to diffuse 1 μm and 1000 s for 1 mm. Therefore, it is crucial to develop efficient micromixers to increase the mixing efficiency for the development of microfluidic systems.

The mixing efficiency is a key parameter for a micromixer. Some methods have been proposed to evaluate the mixing efficiency. A commonly used method is based on the intensity of segregation.

The standard deviation of pixel intensity or point concentration was often used as the mixing index (MI) to evaluate the mixing efficiency [3] and could be expressed by

$$\text{MI} = \sqrt{\frac{1}{N}\sum_{i=1}^{N}(c_i - \bar{c})^2} \tag{1}$$

where, c_i is the point concentration/pixel intensity, $\bar{c}$ is the mean concentration/intensity, and N is the number of sampling points. An improved mixing index [4] based on the comparison of the standard deviation to the mean concentration/intensity was also reported and could be expressed by

$$\text{MI} = 1 - \frac{\sqrt{\frac{1}{N}\sum_{i=1}^{N}(c_i - \bar{c})^2}}{\bar{c}} \tag{2}$$

Besides, the mixing index [5,6] based on the comparison of the standard deviation of the point concentration or pixel intensity in the mixing section to that in the non-mixing section were proposed and could be expressed by

$$\text{MI} = 1 - \frac{\sqrt{\frac{1}{N}\sum_{i=1}^{N}(c_i - \bar{c})^2}}{\sqrt{\frac{1}{N}\sum_{i=1}^{N}(c_0 - \bar{c_0})^2}} \tag{3}$$

where, c_0 is the point concentration or pixel intensity in the non-mixing section and $\bar{c_0}$ is the mean concentration or intensity in non-mixing section. Another mixing index [7] based on the comparison of the integral of the point concentration or pixel intensity in the mixing and non-mixing section was also reported and could be expressed by

$$\text{MI} = 1 - \frac{\int_0^H |c_i - c_\infty| dy}{\int_0^H |c_0 - c_\infty| dy} \tag{4}$$

where, H is the width of the section and c_∞ is the complete mixed concentration (0.5).

Micromixers are often classified as active and passive mixers [8–10]. Active micromixers generally require external energy sources, such as electrical, magnetic, and sound fields, etc., while passive micromixers don't require external energy input except the energy for driving the fluids and often use complex channel geometries to enhance the diffusion or chaotic advection. The structures of active micromixers are often relatively simple and the mixers are easier to control, but the requirement of external energy sources makes them more difficult to integrate. Passive mixers are much easier to integrate into microfluidic devices, but they often require complex fabrication processes.

Although two recent excellent reviews on micromixers have been reported [11,12], one of them summarized both passive and active micromixers that had been developed before 2011, and the other published in 2016 summarized recent passive micromixers. However, in recent years, both active and passive micromixers have been reported based on new principles and structures. Thus, the advances on both active and passive micromixers over the past five years were reviewed in this study.

2. Active Micromixers

Active micromixers depend on different external energy sources to disturb the fluids, increase the contact area, or induce the chaotic advection, thus enhancing the mixing effect. Based on the types of external energy sources, the active micromixers can be further categorized as pressure field driven [7,13–16], electrical field driven [17–33], sound field driven [6,34–43], magnetic field driven [44–57], and thermal field driven [58–63], etc. (Table 1).

Table 1. Active micromixers reported in recent five years.

Energy Source	Characteristic	Mixing Time (s)	*Re*	Mixing Efficiency	Reference
Electrical field	Conductive sidewall	0.033	<1	-	[21] [b]
	Ferrofluid flow	20	-	-	[23] [c]
	Circular copper electrodes	110	-	>90%	[20] [b]
	Asymmetric lateral structure	-	-	100% [2]	[26] [a]
	Floating electrode	1.15	-	95% [3]	[25] [c]
Pressure field	Pulse width modulation	-	83–125	90% [2]	[15] [a]
	Braille Pin Actuator	0.5	4	90% [2]	[14] [b]
	Rotary peristaltic micropump	1	-	90% [4]	[7] [c]
	Single-chamber micropumps	0.45	0.03–30	92% [2]	[16] [a]
Magnetic field	Permanent magnet	80	-	90% [2]	[57] [c]
	Magnetohydrodynamic Actuation	-	0.12	81% [1]	[47] [a]
	Rotating magnetic field	-	-	90–92% [2]	[52] [c]
	Hybrid gradient magnetic field	8	-	97–99% [4]	[48] [a]
	rotating magnetic microbeads	2.5–9	-	-	[54] [c]
Sound field	Bubble cavitation	0.100	0.01	92% [2]	[38] [b]
	Vibrating membrane	0.003	-	90% [3]	[6] [c]
	Bubbles	0.05	0.01	93% [2]	[43] [b]
	Micro-pillars	6	-	-	[40] [b]
	Sharp-edges	0.18	-	-	[36] [b]

[a] Research including only simulated results. [b] Research including only experimental results. [c] Research including both of simulated and experimental results. [1] Mixing Index calculated based on Equation (1). [2] Mixing Index calculated based on Equation (2). [3] Mixing Index calculated based on Equation (3). [4] Mixing Index calculated based on Equation (4).

2.1. Pressure Field Driven Micromixers

The pressure field driven micromixers often have simple structures and consist of a main channel with a side channel (Figure 1a) [64], a main channel with multiple side channels (Figure 1b) [65], or two cross channels (Figure 1c) [66].

The typical pressure field driven micromixer is based on alternate perturbation, which was first reported by Deshmukh et al. [67] using the pulsatile flow micropumps to induce alternate perturbation on fluids in 2000. Some similar micromixers were then reported for mixing two fluids with different flow characteristics and hydrodynamic instability [15,16,68]. One common design of the pulsatile pressure driven micromixer uses two micropumps and a typical T-type channel [16]. These two pumps are used to alternately inject the fluids into the channel. The contact area of the two fluids is greatly enlarged, resulting in a better diffusion and thus a higher mixing efficiency. Additionally, Khoshmanesh et al. [69] presented a simple pressure field driven micromixer with gas bubbles. This micromixer comprises of a main channel filled with water and a side channel connected to a hydrodynamic actuator by a feeder tube to generate bubbles. By oscillating the bubbles at a given frequency, the displacement of bubbles could enhance the mixing efficiency within the main channel.

For the pulsatile pressure driven micromixers, the phase difference of the alternating voltages applied on the two pulsatile pumps has a great impact on the mixing efficiency. Sun and Sie [70] developed a pulsatile pressure driven micromixer with a diverging T-type channel. The phase differences ranging from 0 to π were compared and it was found that the mixing efficiency could reach 95% at the optimal phase difference of 0.5π and the optimal diverging angle of $55°$.

Another typical pressure field driven micromixer is based on oscillatory perturbation, which is generated by extruding and vibrating the side channels [4,7,13,14,71]. Lee et al. [71] presented a typical micromixer with periodic pressure perturbations in the side channels to fold and stretch the main stream. A common pneumatic micromixer is shown in Figure 1d [13], which was comprised of an S-shaped structure with two mixing chambers, two barriers and two pneumatic chambers. With a pumping frequency of 50 Hz, the micromixer could achieve efficient mixing over a wide range of flow rates from 1 μL/min to 650 μL/min. CdS quantum dots were prepared successfully by this micromixer and showed a sharper absorption spectra than those prepared by using the conventional method. Similarly, Tekin et al. [4] used two pairs of chambers on both sides of the main channel to induce an unidirectional flow in the micromixer. By releasing the chambers on one side and at the same time pressurizing the chambers on the other side periodically, complete mixing could be achieved in 350 ms (milliseconds). As shown in Figure 1e, Abbas et al. [14] designed another interesting pressure driven micromixer in a polydimethylsiloxane chip using Braille pin actuation at a resonance frequency (10 Hz) on the side channels to stretch and fold the fluid in the main channel, thus achieving chaotic mixing. This micromixer was successfully used for a continuous dilution of a yeast cell sample by a ratio down to 1:10.

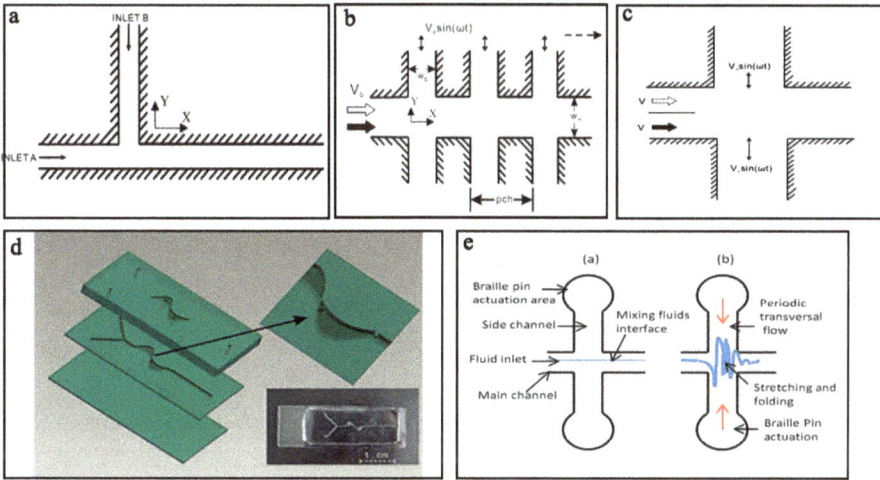

Figure 1. Schematic of pressure field driven micromixers with (**a**) a main channel and a side channel; (**b**) a main channel and multiple side channels; (**c**) two cross channels; (**d**) two mixing chambers, two barriers and two pneumatic chambers; and (**e**) Braille pin actuator. Reproduced with permission from [13,14,64,65].

2.2. Electrical Field Driven Micromixers

The electrical field driven micromixers are mainly based on electro-hydrodynamic (EHD) instability [22], which often uses the motion of electrically charged fluids under an alternating current (AC) or direct current (DC) electric field to disturb the interface of the fluids.

One typical electrical field driven micromixer was presented by Huang [17]. It applied a time-periodic electric field on an electrode array to generate electro-thermal vortices at the corners of each pair of electrodes. The vortices could induce the convective diffusion and thus mix the fluids efficiently. Huang [17] used an AC signal with a peak-to-peak voltage of 6 V, a frequency of 1 MHz, and a phase shift of 180° to actuate an electrode array with an electrode width of 100 μm, an electrode spacing of 30 μm, and an electrode set of 3 in a channel with a width of 400 μm and a height of 30 μm. The mixing efficiency could achieve 94% in ~30 min. Additionally, it was found that the number of the vortices at the corners of the electrodes was reduced from 4 to 2, while the electrode width was the same as the electrode spacing. Zhou et al. [26] employed DC voltage to generate in-plane vortices in an improved microchannel with asymmetric lateral structure and earned a better mixing efficiency.

Electro-kinetics (EKI) is a branch of EHD that describes the coupling of ion transport, fluid flow and electric fields and can be distinguished from EHD by the relevance of interfacial charge at solid-liquid interfaces [33,72]. Electro-kinetic flow instabilities occur under high electric fields in the presence of electrical conductivity gradient [29]. EKI has two basic types of configurations (Figure 2) [72]. For type I, the electric field is orthogonal to the conductivity gradient. For type II, the electric field is parallel with the conductivity gradient, and the net charge density has a non-trivial distribution even in the base state.

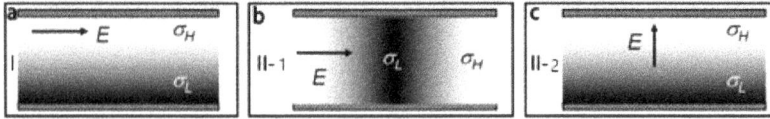

Figure 2. Typical base states for electro-kinetics with type I (**a**); type II-1 (**b**); and type II-2 (**c**). E and the arrow indicate the electric field and its direction, respectively; σ_H and σ_L indicate high- and low-conductivity regions, respectively. Reproduced with permission from [72].

Kumar et al. [23] demonstrated for the first time the electrokinetic instabilities of ferrofluid/water that flowed in a T-shaped channel, and found similar dynamic behaviors in the ferrofluid/water interface at various electric fields. Posner et al. [19] also presented a study on convective electrokinetic instability in a three-inlet, one-outlet electrokinetic focusing flow configuration where the center sample stream and sheath flows had different ionic conductivities (Figure 3). Electrokinetic flowed with conductivity gradients turned unstable when the electroviscous stretching and folding of conductivity interfaces grew faster than the dissipative effect of molecular diffusion. The results showed that the flow became unstable at a critical electric Rayleigh number ($Ra_{e,e} = 205$) for a wide range of conductivity ratios γ (three orders of magnitude) and applied field ratios β.

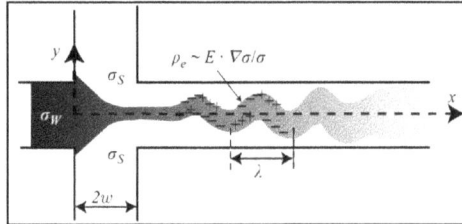

Figure 3. Schematic of the unstable flow in a cross-shaped microchannel with the characteristic D-shape and cross-sections of isotropic etching. σ_s is the ionic conductivity of the sheath streams from top and bottom inlets. σ_w is the ionic conductivity of the sample stream from the left inlet. Pe, E, λ, and w are charge density, electric field, nominal wavelength and half-width of channel, respectively. Reproduced with permission from [19].

Another typical electrical field driven micromixer with liquid metal was described by Tang et al. [73]. Due to their former concept of surface tension driven flow [74], this micromixer used the same Galinstan cap (semi-spherical) placed on a circular copper substrate seat as the core of a liquid metal actuator to induce chaotic advection (Figure 4a). Under a sinusoidal AC electric field (4 V, 50 Hz to 150 Hz), periodic deformation of the Galinstan cap could be observed due to the tangential force, which pulls the surrounding liquid along the surface from the regions of low surface tension (LST) to the regions of high surface tension (HST) (Figure 4b,c). The time averaged mixing efficiency could reach 95% at the flow rate of 25 μL/min when the 4 V & 50 Hz signal was applied.

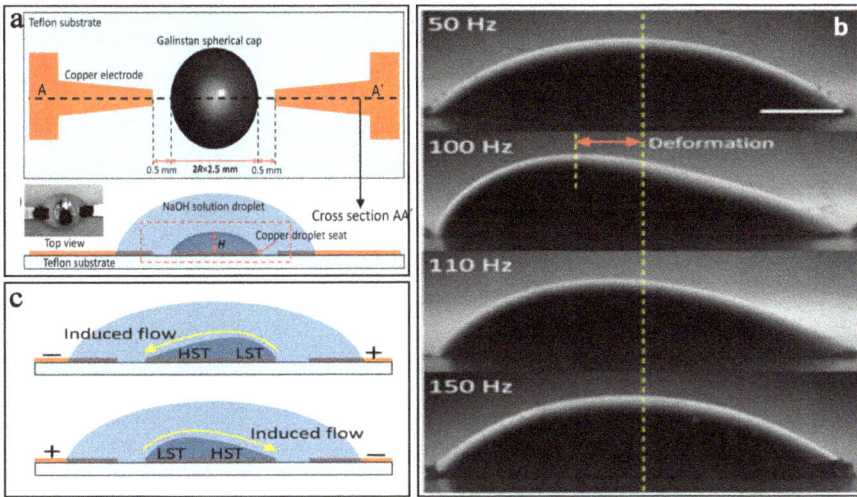

Figure 4. (**a**) Schematic of the micromixer with Galinstan cap; (**b**)The deformation of the Galinstan cap by applying sinusoidal signals with different frequencies and magnitudes; (**c**) Flow velocity vectors (m/s) along the Galinstan surface. Reproduced with permission from [73].

2.3. Sound Field Driven Micromixers

Sound field driven micromixers are based on acoustic resonant disturbance, which was first reported by Moroney et al. [75] using a Lamb-wave membrane device to enhance the mixing.

One typical sound field driven micromixer is based on the use of microbubbles to achieve fast convective mixing [76]. The bubble-based acoustic micromixer with a microstreaming flow field was first reported by Liu et al. [77,78] and it was demonstrated to successfully accelerate the rate of the DNA hybridization process (~5 times faster). However, it generated too many bubbles in the channel. To overcome this drawback, Ahmed et al. [79] developed a single bubble based acoustic micromixer shown in Figure 5a and verified that it could realize complete mixing in 7 ms by trapping the air bubbles in the "horse-shoe" structure to induce microstreaming in the microchannel.

Some other bubble-based sound micromixers have also been reported. Ozcelik et al. [38] utilized the surface roughness of the polydimethylsiloxane (PDMS) microchannel's sidewalls to cavitate the bubbles and obtained excellent mixing efficiency (92%) for high-viscosity fluids at a low Reynolds number of 0.01 in less than 100 ms. Wang et al. [80] presented another bubble based micromixer, which was made up of a 300 μm thick dry adhesive layer sandwiched between two 2-mm-thick polymethylmethacrylate layers. A nozzle-shaped chamber with an acoustic resonator profile was developed in the adhesive layer for generating the bubbles in the microchannel when the piezo-electric disk under the chamber was actuated at the frequency range of 1–5 kHz. Besides, nitrogen gas was reported for generating bubbles to develop bubble based micromixers. As shown in Figure 5b, the nitrogen gas was injected in the center of two reagents in the microchannel and microstreaming was generated for mixing these two reagents [43]. The proposed micromixer could mix two highly viscous fluids (95.9 mPa·s) in the presence of an acoustic field within 50 ms with an excellent mixing efficiency of ~93% at a low *Re* number (~0.01).

Another typical sound field driven micromixer is based on a surface acoustic wave (SAW), which is an acoustic wave traveling along the surface of a solid material [81]. As shown in Figure 5c, Luong et al. [82] reported the use of focusing interdigitated electrodes instead of traditional parallel interdigitated electrodes to concentrate the acoustic energy. The SAW was generated by the interdigitated electrodes deposited on the piezoelectric substrate to induce mixing due to the disturbance of the transversal acoustic streaming. The mixing efficiency of 90% was obtained with the peak-to-peak voltage of 80 V and the Peclet number of 74.4×10^3. Besides, vibrating membrane, micro-pillars and sidewall sharp-edge were also reported in the development of the continuous-flow micromixers. As shown in Figure 5d, Phan et al. [6] developed a vibrating membrane with a hole to generate strong streaming vortices in the channel. Besides, micro-pillars was also utilized to realize homogeneous mixing in 6 s by Oever et al. [40] in a centimeter-scale acoustic micromixer. Huang et al. [36] reported the oscillation of sidewall sharp-edges to induce an acoustic streaming to achieve excellent mixing in 180 ms (Figure 5e).

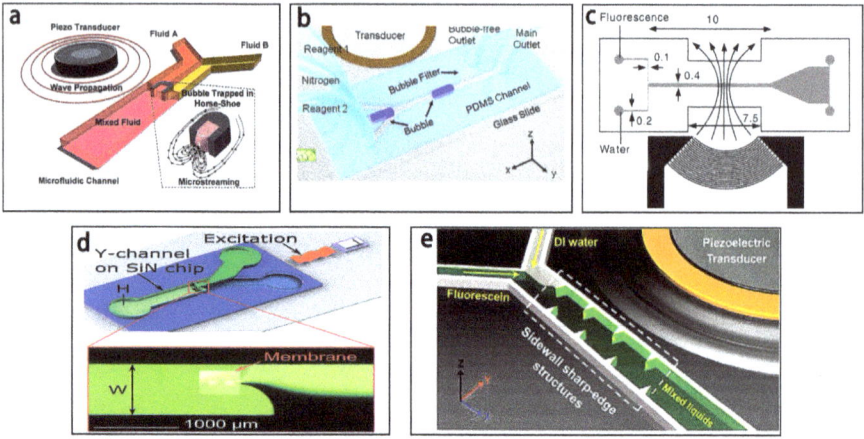

Figure 5. Schematic of the sound field driven with (**a**) "horse-shoe" structure; (**b**) nitrogen gas; (**c**) interdigitated electrodes; (**d**) vibrating membrane; and (**e**) sidewall sharp-edges. Reproduced with permission from [6,36,43,79,82].

2.4. Magnetic Field Driven Micromixers

Magnetic field driven micromixers are mainly based on magneto-hydrodynamics (MHD) and magnetic stirring.

Magneto-Hydrodynamic micromixers often utilize AC or DC electric fields and magnetic fields to apply Lorentz forces on the magneto-fluids, which can induce secondary flows for stirring and mixing. One typical magneto-hydrodynamic micromixer [83] is shown in Figure 6a, which consisted of a conduit filled with an electrolyte solution, and individually controlled electrodes patterned along its double sidewalls. When the micromixer was placed in a uniform magnetic field, it could serve as both a mixer and a pump. Recently, ferrofluid was extensively employed for the studies on magnetic micromixers [48,49,53,56,57]. A ferrofluid-based microfluidic magnetic micromixer developed by Cao et al. [48] using a hybrid magnetic field generated by some micro-magnets and an external AC uniform magnetic field to apply periodic magnetic forces on the ferro-fluid, thus achieving a high mixing efficiency (97%) in 8 s at a distance of 600 µm from the mixing channel inlet. As shown in the Figure 6b is another ferrofluid-based magnetic mixer developed by Nouri et al. [57] using a Y-shaped microchannel with a permanent magnet to mix deionized water and Fe_3O_4 ferrofluid. The ferrofluid migrated from one bottom side of the channel to the top side under the magnetic field generated by the permanent magnet, resulting in the mixing of the two fluids.

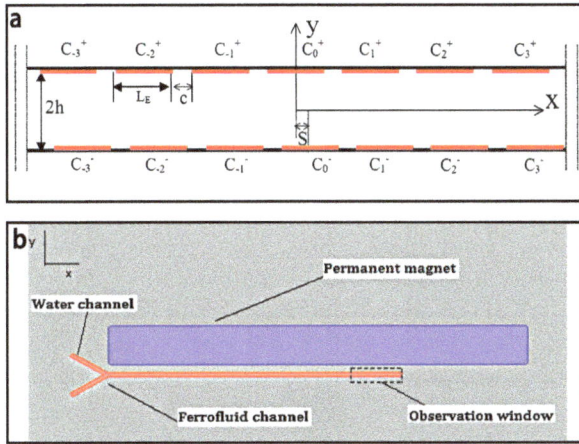

Figure 6. (**a**) Schematic of the magneto-hydrodynamic micromixer. L_E is the width of the electrodes, c is the width of the space between two adjacent electrodes, S is the offset of two facing elecrodes, and $2h$ is the width of the conduit; (**b**) Schematic of the ferrofluid-based magnetic micromixer with Y-shaped channel. Reproduced with permission from [57,83].

Magnetic stirring micromixers often used magnetic stirrers driven by external rotating magnetic fields to mix the fluids in the chamber. A typical magnetic micromixer is shown in Figure 7a [84]. The millimeter-sized magnetic stirrer was controlled by the rotating field to stir the fluids in the channel and the fluids could reach total mixing within seconds. Veldurthi et al. [50] studied the simulation of this stirring micromixer at different rotating speeds of the magnetic stirrer over a wide range of flow rates, and the results showed that the maximum mixing efficiency (~90%) was achieved at 1500 rpm. Additionally, the rifampicin drug was successfully loaded on TiO_2 nanoparticles by this micromixer. Microbeads were also used as stirrers in microchannels [45,51,54]. Owen et al. [54] proposed a micromixer with short channel lengths (270 μm) by an array of rotating magnetic microbeads (Dynabeads M-280, 2.8 μm diameter), which was attracted to the poles of the NiFe feature driven by an external magnetic field (Figure 7b,c). Complete mixing could achieve in 2.5 to 9 s depending on different flow velocity and specific biological particles in the fluid could be captured by the magnetic microbeads with different functional groups.

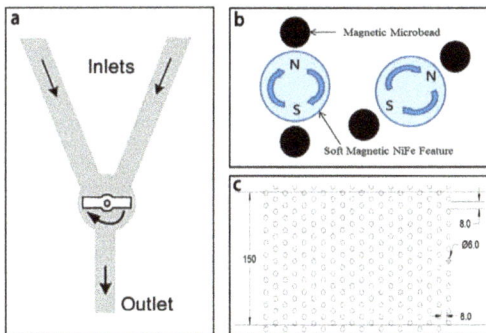

Figure 7. (**a**) Schematic of the magnetic stirring micromixer; (**b**) Schematic of the magnetic microbeads attracted to the poles of the NiFe feature; (**c**) Schematic of the microfluidic channel with its floor patterned with NiFe features. Reproduced with permission from [54,84].

In addition, artificial cilia with embedded magnetic particles driven by a homogeneous magnetic field was used in a simple T-shape channel to realize the mixing of two highly viscous fluids [85]. In this micromixer, a high mixing efficiency of 86% was achieved when the figure-of-eight trajectory of artificial cilia was generated by three rolls of magnetic coils.

2.5. Thermal Field Driven Micromixers

Thermal field driven micromixers are often dependent on the use of thermal bubbles for mixing. Huang [61,86] presented a thermal bubble actuated microfluidic chip with microvalve, micropump and micromixer, based on a simple process with silicon-onisolation (SOI) wafer (Figure 8). The size of thermal bubbles can be controlled at flow rate of less than 4.5 µL/s. When an AC signal at high frequency was applied to the micro-heater, the thermal bubbles could grow periodically and collapse rapidly, thus generating turbulent flow in the fluids and increased mixing efficiency.

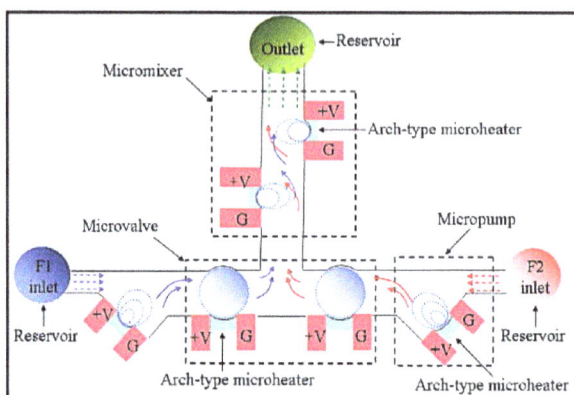

Figure 8. Schematic of the microfluidic system including microchannel, micro-valve, micro-pump and micromixer. Reproduced with permission from [61].

Micromixer driven by electrothermal effect also involves lots of multi-physics phenomena which could be used in micromixers [59,60,62]. Recently, Kunti et al. [63] proposed an alternating current electrothermal micromixer consisted of eight pairs of asymmetric electrodes with AC voltage. As shown in Figure 9, in this micromixer waviness of the floor increased the contact area between two fluids and lateral vortex pairs were generated by symmetric electrode pairs located on the top wall. A mixing efficiency of 97.25% can be achieved under a flow rate of 1.794 µm^2/min.

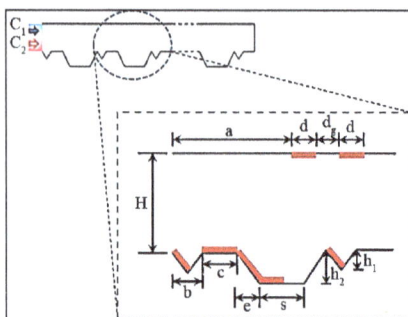

Figure 9. Schematic of the alternating current electrothermal micromixer. Reproduced with permission from [63].

2.6. Other Field Driven Micromixers

Centrifugal forces can also enhance mixing and has been used in lots of micromixers [87–89]. Haeberle et al. [90] reported a centrifugal micromixer, relying on the Coriolis force induced by the rotated plate to drive and mix the fluids (Figure 10a). Base on the centrifugal micromixer, Leung et al. [91] investigated the mixing efficiency of different rotating radial microchannels with various obstructions and/or width-constriction geometries (Figure 10b–e). The experimental results showed that transverse flow in the microchannel was highly increased due to the obstruction with constriction (OWC) configuration, local centrifugal acceleration, and Coriolis acceleration. Moreover, for the rotating OWC (obstruction follow by width-constriction) channel, the mixing efficiency could reach 95% at the distance of 30 mm from inlet when the rotating rate was 73 rad/s, which was much more than those of the stationary OWC channel, the rotating unobstructed/obstructed channel, and the rotating width-constricted channel.

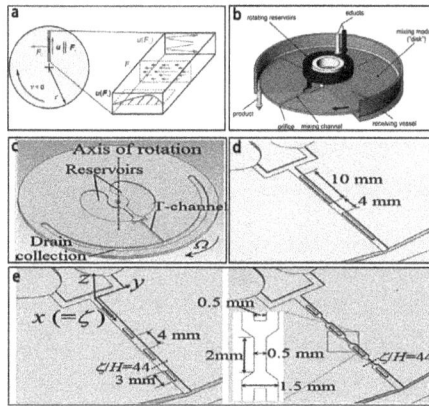

Figure 10. (**a**) The force analysis: F_c is the Coriolis force, F_v is the centrifugal force, $u(F_c)$ is the transverse of fluids; (**b**) Schematic of the centrifugal mixer; (**c**) Schematic of the microchannels; (**d**) with two obstructions; and (**e**) with four obstructions. Reproduced with permission from [90,91].

3. Passive Micromixer

Passive micromixers—also called static micromixers—are based on the structure of the microchannels to enhance molecular diffusion and chaotic advection for efficient mixing [92]. There is an excellent 2004 review focused on passive micromixers by Nguyen et al. [8], yet there are many new passive micromixers developed recently by scientists. According to the dimensions of the structure, passive micromixers can be sub-classified as either three-dimensional (3D) and two-dimensional (2D). Over the past five years—as shown in Table 2—many new passive micromixers based on the structure of T-type [93], Zigzag [94], and Serpentine [3], etc., have been reported.

Table 2. Active micromixers reported in recent five years.

Dimension	Structure	Characteristic	*Re*	Mixing Length (µm)	Mixing Efficiency	Reference
2D	Unbalanced collisions channel	Unbalanced three-split recombine sub-channels	30–80	8275	90% [3]	[95] [a]
		Dislocation structure	<80	8000	85% [2]	[96] [c]
	Embedded Barriers channel	Triangle baffle	1	6400	85.5% [3]	[97] [c]
		Curved micromixers with cylindrical obstructions	0.1–60	8280	88% [3]	[98] [a]
	Spiral	Single logarithmic spiral	67	12,000	86% [4]	[99] [c]
		Double logarithmic spirals	50	5000	80% [3]	[100] [a]
	Convergent–divergent channel	Sigma channel	0.91	8000	79.1% [3]	[101] [a]
		Semi-elliptical walls	35.5	-	80% [2]	[102] [c]
		Convergent–divergent walls	10–70	6720	90% [3]	[103] [a]
		Ellipse-like micro-pillars	≤1	9000	80% [2]	[104] [c]
		Reversed flow in square wave channel	≤0.1 or ≥10	3710	95% [2]	[105] [a]
		Reversed flow in zigzag channel	≤0.5 or ≥5	-	93% [2]	[106] [a]
3D	Chamber	Trapezoidal chambers	0.5–60	3870	80% [2]	[107] [a]
		Trapezoidal-zigzag channels	0.1–0.9 or 20–80	3610	90% [3]	[108] [a]
		Unbalanced split and cross-collision chambers	0.5–100	5000	80% [2]	[109] [c]
		Circular mixing chambers	0.1	6400	88% [3]	[110] [a]
		Split and recombine chambers	1–100	-	90% [3]	[111] [c]
	3D Spiral	Three dimensional spirals	40	2340	90% [1]	[112] [c]
		Cross-linked dual helicals	0.003–30	320	99% [2]	[113] [a]
		Tapered structures	50	10,500	90% [3]	[114] [c]
	Overbridge	Overbridge-shaped channel	0.01–50	2000	90% [4]	[115] [c]
		Tesla structures	0.1–100	10,700	94% [2]	[116] [c]
		X-shape structures combined with H-shape structures	0.3–60	102,500	87.7% [3]	[117] [c]
		X-shape structures combined with O-shape structures	0.3–60	102,500	72.9% [3]	[117] [c]
		Serpentine crossing channels	0.2–10	7500	99% [3]	[118] [c]

[a] Research including only simulated results. [b] Research including only experimental results. [c] Research including both of simulated and experimental results. [1] Mixing Index calculated based on Equation (1). [2] Mixing Index calculated based on Equation (2). [3] Mixing Index calculated based on Equation (3). [4] Mixing Index calculated based on Equation (4).

3.1. 2D Passive Micromixers

2D passive micromixers with simple planar structures such as obstacles, unbalanced collisions, convergence–divergence channels, and spiral channels etc. are easy to fabricate with lithography method and generate chaotic advection due to the special shape of the channel.

3.1.1. Obstacle Based Micromixers

The obstacle based micromixers are mostly combined with various embedded grooves or barriers with different shapes and heights. One typical obstacle based micromixer with straight grooves was first proposed by Stroock et al. [119]. The experimental results showed that these straight grooves aroused a secondary flow in the channel and a good mixing over a wide range of Reynolds numbers (0 < *Re* < 100). Howell et al. [120] improved this design by placing grooves in both the top and bottom of the channel. Besides, Hossain et al. [121] further optimized this micromixer, and simulation results showed that the best mixing efficiency could reach 91.7%.

Another typical obstacle for these micromixers is based on the barriers in the channels. Bhagat et al. [122] studied the effect of the barriers' height and shape on the mixing efficiency.

The simulation results showed that the mixing efficiency increased when the higher barriers were used. Specifically, when the barriers had the same height as the channel the mixer was called as a split-and-recombine (SAR) one and had the best mixing efficiency. For the stepped-diamond-shaped barriers, the efficiency could reach 77%. Some similar researches [123,124] also proposed the numerical and experimental investigation on comparing the mixing behaviors of microchannel that was shaped with various kinds of barriers. They showed that by increasing the number and length of rectangular barriers can potentially enhance the mixing effect within a short mixing length in microchannels. As shown in Figure 11a, Wang et al. [97] fabricated a passive micromixer containing 64 groups of triangle barriers with excellent mixing efficiency for Reynolds numbers in the range of 0.1 to 500. Both simulation and experimental results showed that the bigger apical angles and the more groups of the triangle barriers led to the better mixing efficiency and the best mixing efficiency could reach 91.2% (Figure 11d).

Many studies [125–127] have demonstrated that the curved channel based micromixer, without obstacles, could not achieve a high mixing efficiency unless it had a high Reynolds number. To improve the mixing efficiency of the curved channel based micromixer at the low Reynolds number, Tsai et al. [128] proposed a planar micromixer based on multidirectional vortices in the curved channel with two radial barriers of 40 μm thick and 97.5 μm long (Figure 11b). The effects of the position and size of the radial barriers were studied (Figure 11e), and it was found that the presence of the Dean vortices [129] generated by the curved channel and the expansion vortices produced by the barriers led to a fine mixing efficiency of ~72% at $Re = 81$ in a very short length (~4.25 mm). Different from the above-mentioned curved channel based micromixers, Afroz Alam et al. [98] presented a new one employing several cylindrical barriers in the curved microchannel (Figure 11c). The barriers in the curved microchannel could generate secondary flows and SAR flows, resulting in a high mixing efficiency of 88% at both low and high Reynolds numbers ($Re = 0.1$ and $15 \leq Re \leq 60$).

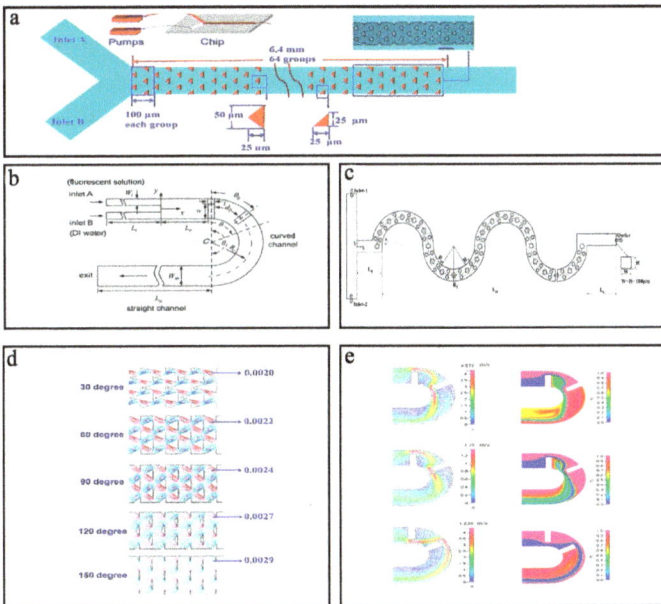

Figure 11. Schematic of the obstacle based micromixer (**a**) with triangle barriers, (**b**) with radial barriers, and (**c**) with cylindrical barriers; (**d**) The simulation results of the obstacle based micromixer with triangle barriers; (**e**) The simulation results of the obstacle based micromixer with radial barriers. Reproduced with permission from [97,98,128].

3.1.2. Unbalanced Collision Based Micromixers

The unbalanced collision based micromixers are often dependent on the asymmetric structure of the channel or the different flow rate of the fluids. One typical unbalanced collision based micromixer was shown in Figure 12a and based on the concept of unbalanced splits and cross-collisions of the fluids, which was first presented by Ansari et al. [130]. The mixing was mainly due to the combined effect of unbalanced collisions and Dean vortices. When $w1/w2 = 2.0$, the best mixing efficiency of 65% for Reynolds numbers ranging from 10 to 80 could be obtained. As shown in Figure 12b, Xia et al. [131] developed an unbalanced circular micromixer using fan-shaped cavities in the major sub-channel to generate convergent-divergent structures with the mixing efficiency of 78%. The stagger structure was used at the corner of the major sub-channel in the previously mentioned mixer with a higher mixing efficiency of 86% (Figure 12c) [96].

Moreover, a similar micromixer with three unbalanced rhombic sub-channels was developed by Hossain [95] and the simulation results showed that the mixing efficiency of the unbalanced three-split rhombus based micromixer (86%) had ~1.44 times than that of the two-split rhombus based one at the Reynolds number of 60 (Figure 12d).

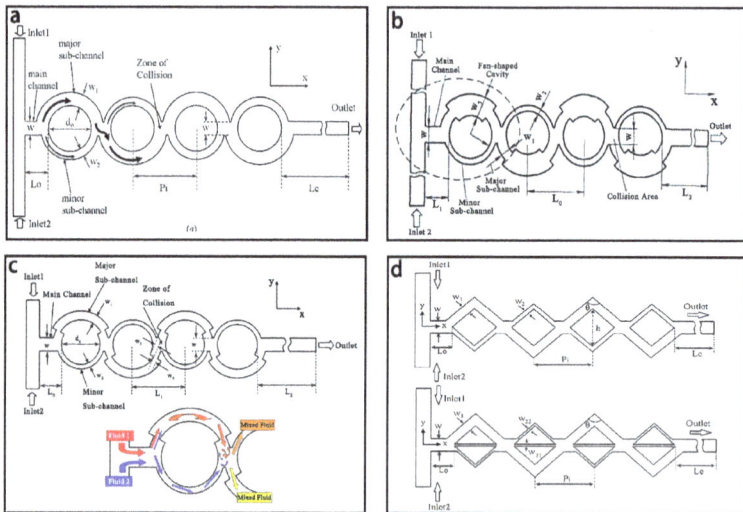

Figure 12. Schematic of the (**a**) unbalanced collision based micromixer; (**b**) unbalanced collision based micromixer with fan-shaped cavity in major sub-channel; (**c**) unbalanced collision based micromixer with stagger structure; (**d**) unbalanced collision based micromixers with two-split rhombus channel and with three-split rhombus channel. Reproduced with permission from [95,96,130,131].

3.1.3. Spiral Based Micromixers

The spiral based micromixer was first put forward by Schönfeld et al. [132] and is shown in Figure 13a. Subsequently, Sheu et al. [114] combined this typical spiral based micromixer with unbalanced collisions to develop a parallel laminar micromixer with two-dimensional staggered curved channels (Figure 13b). Dean vortices were formed in curved channels by centrifugal forces, and the split structures of the tapered channels resulted in the unbalanced split of the main stream and the reduction of the diffusion distance of two fluids.

Another typical spiral based micromixer is shown in Figure 13c with a maximum mixing efficiency of 86% at $Re = 67$ [99], which is significantly higher than that of the Archimedes and Meandering-S spiral based ones. Similarly, He et al. [100] reported a two logarithmic spiral based micromixer with the spiral polar angle from 0° to 180°. Because the two logarithmic spirals with the variable curvatures were parallel, secondary flows were generated to enhance mixing with the mixing efficiency of 80% at the Reynolds number of 0.2.

An interesting double spiral based micromixer was first proposed by Sudarsan and Ugaz in 2004 [133] and is shown in Figure 13d. The mixing efficiency could reach >90% at the end of the second section. Furthermore, another interesting labyrinth-like multiple spiral based micromixer was shown in Figure 13e and a fast complete mixing within 9.8 s to 32 ms could be achieved for Reynolds numbers between 2.5 and 30 [134]. Similarly, Al-Halhouli et al. [135] presented two spiral based micromixers with interlocking semicircles and omega-shaped channels respectively (Figure 13f,g). Nearly complete mixing could be achieved for a wide range of *Re* between 0.01 and 50 in each micromixer.

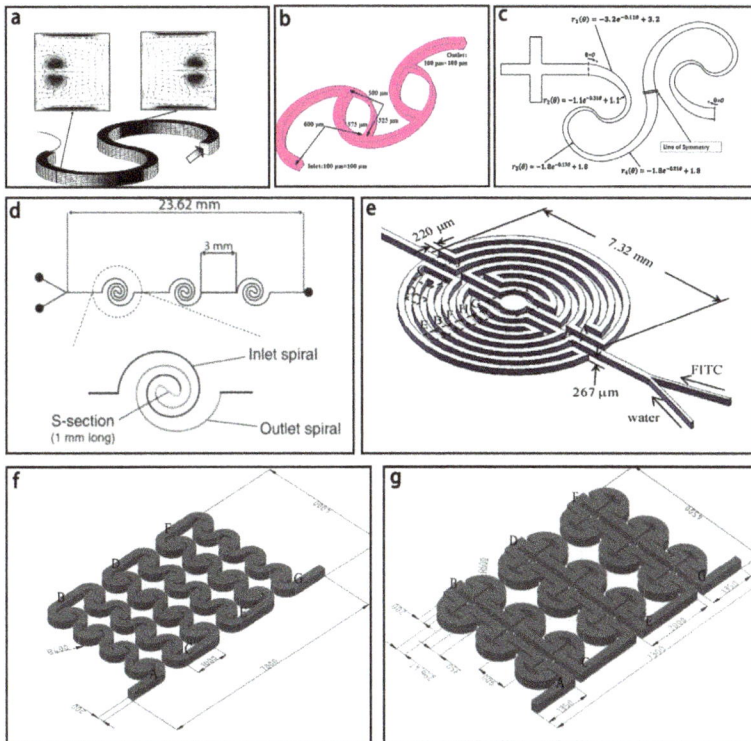

Figure 13. (**a**) Schematic of the spiral micromixer (The arrow indicates the flow direction); (**b**) Schematic of the parallel laminar micromixer; (**c**) Schematic of the spiral based micromixer with sequential logarithmic structure; (**d**) Schematic of the double spiral based micromixer; (**e**) Schematic of the labyrinth-like multiple spiral based micromixer; (**f**) Schematic of the spiral based micromixers with interlocking semicircles; (**g**) Schematic of the spiral based micromixers with omega-shaped channels. Reproduced with permission from [114,132–135].

3.1.4. Convergence–Divergence Based Micromixer

A convergence-divergence structure of micromixers can cause expansion vortices, subsequently causing a great disturbance in the microchannel laminar flow as well as increasing the contact area between the different fluids, thereby enhancing the mixing efficiency.

As shown in Figure 14a, one typical convergence-divergence based micromixer with sinusoidal walls was represented by Afzal and Kim [136]. Coupled with pulsatile flow (Figure 14b), this micromixer could achieve a mixing efficiency of 92% within two periods of the sinusoidal walls. A multi-objective optimization [101] of the Sigma micromixer [137] was proposed. Lengths of the major axis (*a*), minor axis (*b*) and the constirction width (*h*) were optimized at the *Re* = 0.91. The results showed the mixing efficiency increased with higher a/H values and lower values of b/g and h/H and a best mixing efficiency (79.1%) was reached with a/H = 0.75, b/g = 0.503 and h/H = 0.216. Another convergence-divergence based micromixer with meandering channel, presented by Wu and Tsai [102], showed a better mixing efficiency (80% at *Re* = 35.5) than the Sigma micromixer. Different expansion ratios, defined as $E = W_{max}(s)/W_{min}(s)$, were studied and it showed that such micromixer with a larger expansion ratio earned a better mixing (Figure 14c).

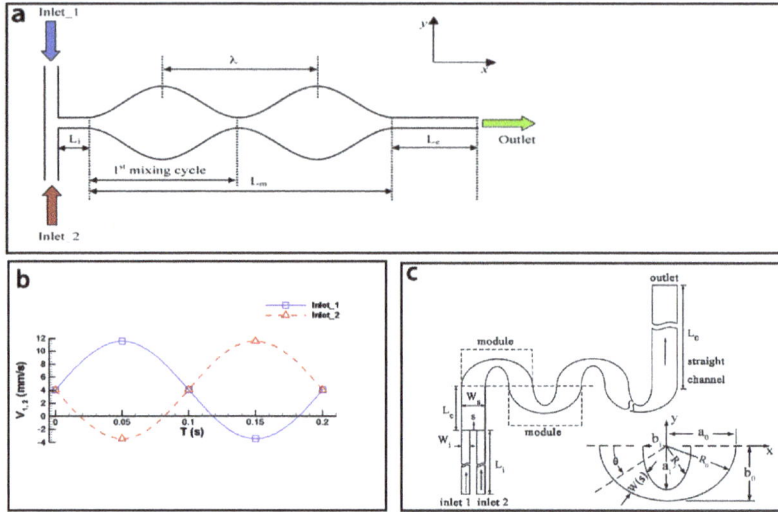

Figure 14. (**a**) Schematic of the convergence-divergence based micromixer; (**b**) The inlet flow velocities of the convergence-divergence based micromixer; (**c**) Schematic of the convergence-divergence based micromixer with meandering channel. Reproduced with permission from [102,136].

Afzal et al. [103] combined the split-and-recombine structure with the convergence-divergence walls to generate secondary flows (Figure 15a). The results showed a decent mixing efficiency of 95% could be achieved with Reynolds numbers ranging from 10 to 70. Tran-Minh et al. [104] proposed a combination of the planar split-and-recombine structure with the ellipse-like micro-pillars and it was successfully used for continuous mixing of human blood (Figure 15b). The optimal parameters (a_1:a_2:b = 5:6:4) for the ellipse were investigated with a high mixing efficiency of >80%, which was better than that of the T-channel mixer (Figure 15c). Recently, several convergence–divergence based micromixers, which were transformed from the typical two-dimensional serpentine channel and based on topology optimization method, were reported by Chen and Li [105]. In these micromixers, convergence-divergence structures were set at the center of the channel with the obstacles at different height (Figure 15d), The results showed that the micromixer with the ratio of the height of the convergence-divergence structure to that of the channel of 0.75 had the best mixing efficiency of over 95% for a wide range of *Re* (*Re* ≥ 5 or *Re* ≤ 0.5). In addition, they used the zigzag channel based on topology optimization to replace the serpentine one to develop a new micromixer [106], and it owned a mixing efficiency of over 93% for a wide range of *Re* (*Re* ≥ 5 or *Re* ≤ 0.5).

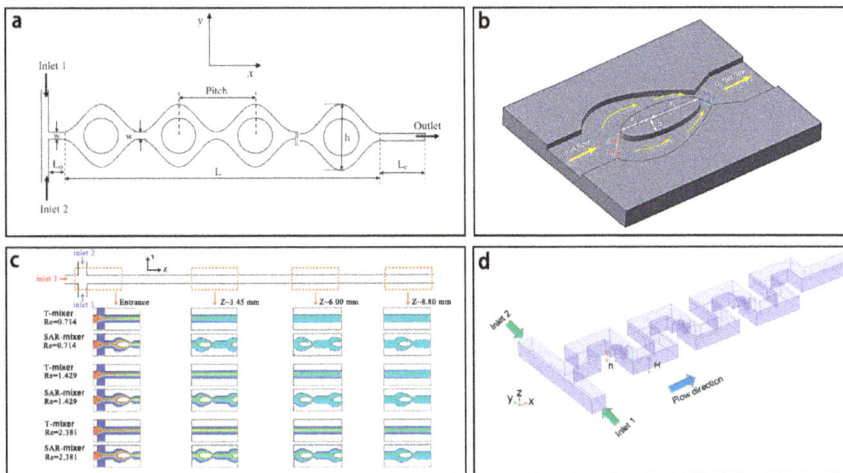

Figure 15. (**a**) Schematic of the convergence-divergence based micromixer with split-and-recombine structure; (**b**) Schematic of the convergence-divergence based micromixers with the ellipse-like micro-pillars; (**c**) Comparison of the mixing between the ellipse-like micro-pillars mixer and the T-channel mixer; (**d**) Schematic of the convergence-divergence based micromixers with two-dimensional serpentine channel and based on topology optimization method. Reproduced with permission from [103–105].

3.2. 3D Passive Micromixers

3D passive micromixers are often dependent on complex spatial structures, which require cumbersome fabrication and can generate various vortices such as second flow vortices, Dean vortices, and chaotic advection etc., to enhance mixing.

3.2.1. Lamination Based Micromixers

Lamination based micromixers usually comprise multilayer structures, and can achieve excellent mixing in milliseconds. One typical lamination based micromixer was first reported by Branebjerg et al. [138]. As shown in Figure 16, Buchegger et al. [139] presented an aclinic multi-lamination based micromixer with wedge shaped vertical fluid inlets for fast and efficient mixing. In this micromixer, two ports with 10 μm width were split into four vertical inlets through the distribution network. Then, 4 fluid layers were formed in the mixing channel to increase the contact area. The simulation results showed that the mixing efficiency could reach 90% in 0.64 ms under a diffusion coefficient of 2×10^{-9} m^2/s. Proton exchange reaction of H_2O and D_2O forming 2 HDO was well achieved. Similarly, SadAbadi et al. [140] designed a simple 3-layer micromixer with a high mixing efficiency of 85% for *Re* < 5.5.

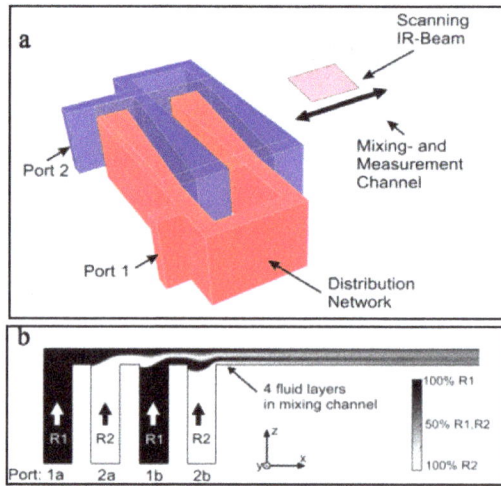

Figure 16. (**a**) Schematic of the lamination based micromixer with four lamination layers; (**b**) Two dimensional flow simulation of the micromixer. Reproduced with permission from [139].

Lim et al. [141] proposed another lamination based micromixer, also called a crossing manifold micromixer (CMM) (Figure 17). It had a three-dimensional microstructure with a sequential configuration of horizontally and vertically crossing tube bundles. In this micromixer, two fluids were rearranged alternately in vertical and horizontal direction for fast mixing by two kinds of mixing modules: horizontally crossing manifold micromixer (H-CMM) and vertically crossing manifold micromixer (V-CMM), respectively. According to the simulation, when V-CMM was set at the distance of 50 mm from H-CMM, the mixing efficiency of 90% could be estimated in the channel length of 250 mm with a total flow rate of 0.003 mL/min.

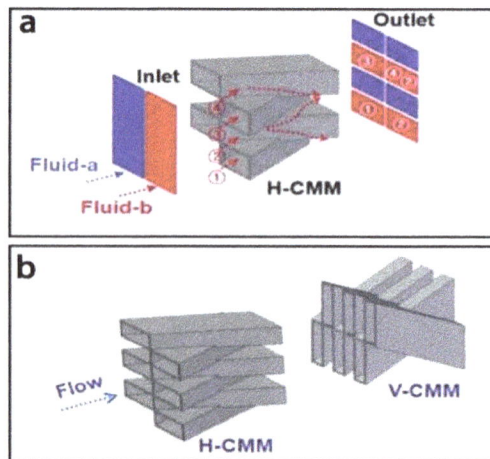

Figure 17. (**a**) Schematic of the lamination based micromixer with crossing manifold structure; (**b**) The horizontally crossing manifold micromixer (H-CMM) and the vertically crossing manifold micromixers (H/V-CMM). Reproduced with permission from [141].

3.2.2. Chamber Based Micromixers

Chambers with special structures, such as a convergence–divergence structure, recirculation structure and counterflow structure, are often used to significantly improve the mixing efficiency in passive micromixers.

One typical chamber based micromixer, using a chamber with convergence-divergence structure was proposed by Hai et al. [107]. Based on the effect of stretching-folding in both vertical and horizontal directions, this convergence-divergence structure was designed as trapezoidal shape to provide a high mixing efficiency for low flowrate fluids. Simulation results showed that it retained a high mixing efficiency of over 80% for a low Reynolds number of between 0.5 and 60 with a total mixing length of 3870 μm. Combining this micromixer with the unbalanced splits and cross-collisions of fluids, an improved micromixer with shifted trapezoidal chambers and a total mixing length of 5000 μm was then presented and shown in Figure 18a [109]. Simulation results showed that the mixing efficiency was over 80% for an entire range of Reynolds numbers from 0.5 to 100. Recently, a novel SAR micromixer, namely the H-C mixer, combining both the H mixer [142] and the Chain mixer [143] (Figure 18c,d), was developed by Viktorov et al. [111] (Figure 18b). Fluid folding, rotation and expansion occurred in the channel with the splitting-recombination and convergence–divergence structures, thus resulting in a good mixing efficiency of over 93%. The H-C mixer could be considered for industrial applications due to its simple manufacturing procedure, great mixing efficiency and low pressure drop.

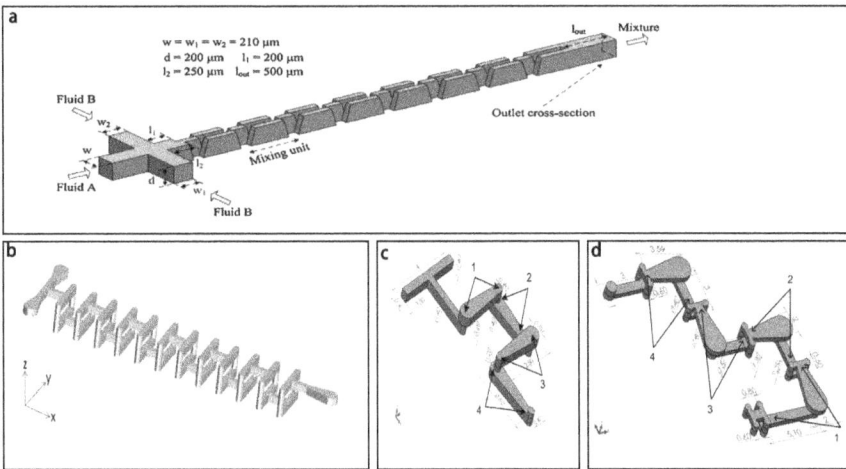

Figure 18. (**a**) Schematic of the chamber based micromixer with shifted trapezoidal chambers; (**b**) Schematic of the H micromixer; (**c**) Schematic of the chain micromixer; (**d**) Schematic of the H-C micromixer. Reproduced with permission from [111,144].

Another typical chamber based micromixer with circular chambers has already been reported [145–147] and circular chambers have been found to be effective in mixing over a wide range of Reynolds numbers. Alam et al. [110] proposed a chamber based micromixer with eight circular chambers and two constriction channels to connect the adjacent chambers (Figure 19). Simulation results showed that this micromixer could achieve a mixing efficiency of 88% at a low Reynolds number (*Re* = 0.1) where diffusion dominated the fluidic mixing.

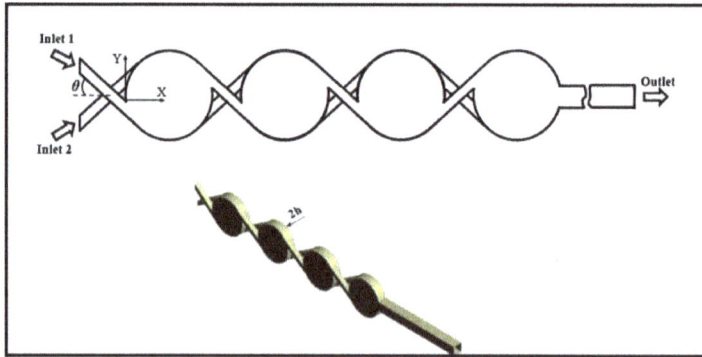

Figure 19. Schematic of the chamber based micromixers with circular chambers. Reproduced with permission from [110].

3.2.3. 3D Spiral Based Micromixers

One typical 3D spiral based micromixer with two spiral microchannels and an erect channel was presented by Yang [112] (Figure 20a). For *Re* > 40, the maximum mixing efficiency could be up to 90% and the erect channel played a significant role in mixing.

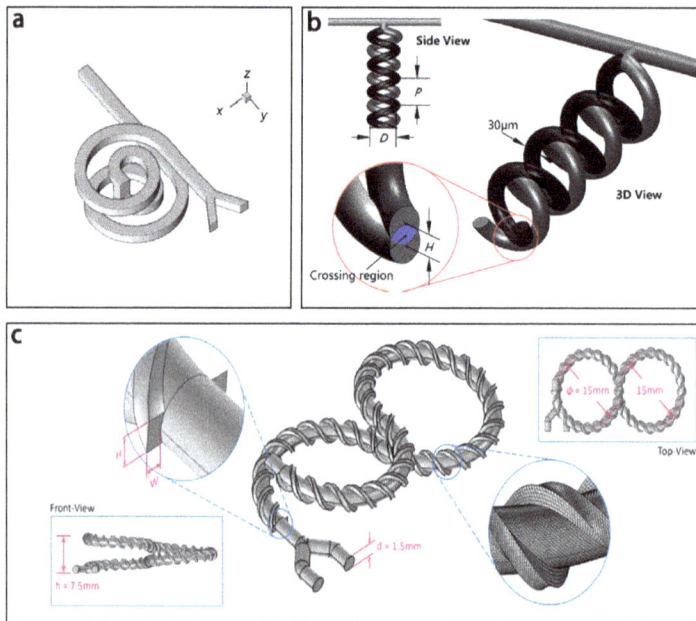

Figure 20. Schematic of the 3D spiral based micromixer (**a**) with two spiral microchannels and an erect channel; (**b**) with double helical channels in opposite directions; and (**c**) with 3D spiral and fine-threaded microchannel. Reproduced with permission from [112,113,148].

Recently, Liu et al. [113] proposed a novel 3D spiral based micromixer (Figure 20b), which consists of double helical channels in opposite directions to create repeated crossing regions. The simulation and experimental results showed that the micromixer had a high mixing efficiency of 99% for a wide range of low *Re* (0.003–30). In addition, Rafeie et al. [148] presented an effective micromixer (mixing efficiency > 90%) which incorporates the 3D spiral and fine-threaded microchannels for a wider range of *Re* (1–1000) (Figure 20c).

3.2.4. Overbridge Based Micromixers

Overbridge based micromixers, often with 3D structures connected by a bridge-shaped channel, are mainly based on the concept of splitting and recombination.

One typical overbridge based micromixer with splitting channels of unequal widths was proposed by Li et al. [115] and it had a high mixing efficiency of over 90% for a *Re* range from 0.01 to 200 according to the simulation results and the same mixing efficiency for a *Re* range from 0.01 to 50 by the experimental results. And the mixing efficiencies with different inlet flow rates ranging from a ratio (fluid 1: fluid 2) of 1:9 to 9:1 were also compared and it showed that the best mixing efficiency (100%) was reached with the ratio of 1:9 at *Re* = 0.01. Feng et al. [117] used the X-shaped structures to connect the O-shaped structures or H-shaped structures in passive micromixers with total length of 10.25 mm (Figure 21a,b). Experimental results showed that the mixing efficiency of the micromixer with X-shaped structures and H-shaped structures and the micromixer with X-shaped structures and O-shaped structures were from 91.8% to 87.7% and 89.4% to 72.9% respectively when the *Re* increases from 0.3 to 60.

Based on the well-known Tesla structures, another overbridge based micromixer was fabricated by Yang et al. [116] (Figure 21c). Binding reactions between the antibodies and the antigens for the detection of cancer cells were efficiently realized by this micromixer. By placing Tesla structures on other Tesla structures, this micromixer could realize a mixing efficiency of 94% for a *Re* range from 0.1 to 100. Simulation results also showed that a larger contact area between two Tesla structures led to better mixing.

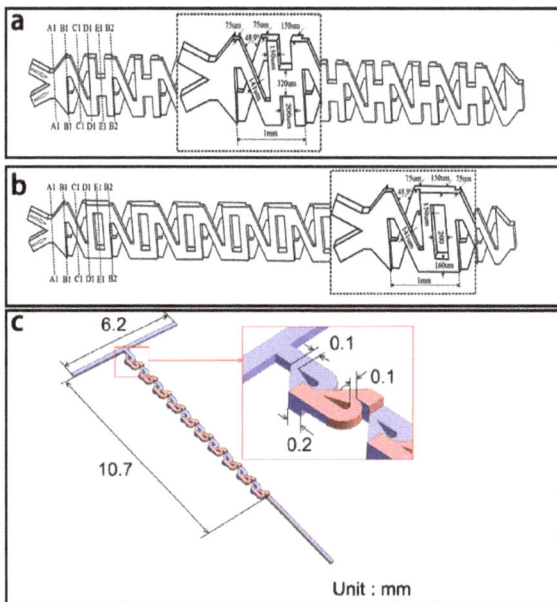

Figure 21. Schematic of the overbridge based micromixer (**a**) with XH-shaped structures; (**b**) with XO-shaped structures; and (**c**) with Tesla structures. Reproduced with permission from [116,117].

4. Conclusion and Future Trends

Micromixing has made rapid developments over the past decade due to advances in MEMS and Microfluidics. Compared with conventional macro-scale mixers, both passive and active micromixers have demonstrated their features of faster mixing, easier fabrication, higher efficiency, and lower cost. This paper systematically reviewed the recent advances in various active and passive micromixers. The micromixers with different structures and external fields were discussed and their advances and defects were also pointed out for reference.

With the increasing needs from the biomedical, agricultural, food and environmental fields, microfluidic chips for rapid and automatic screening or monitoring of such biological and chemical targets as glucose, pathogens and melamine, etc. have attracted more and more attention, which require the micromixing technologies to boost the on-chip biochemical detection assays. Thus, the integration of micromixers with biochemical sensors will be a promising trend. In recent years, 3D printing technology—with an accuracy of up to several micrometers and with the incorporation of various materials—has been widely used for the development of complex structures and the fabrication of various valves and pumps in a very short time [149]. The use of 3D printing technology to fabricate micromixers with complex structures more accurately, easily and faster at lower cost is very promising. Additionally, paper-channel micromixers combined with external fields have the potential to provide simple, low-cost and disposable methods for point-of-care diagnostics.

Acknowledgments: This research is supported by the National Key Research and Development of China (2016YFD0500706) and the Yunan Academician Expert Workstation (Wang Maohua, Grant No. 2015IC16).

Author Contributions: G.C. performed the document indexing and manuscript writing. L.X. and H.Z. performed the figure processing and data analysis. J.L. proposed and designed the whole manuscript.

Conflicts of Interest: The authors declare no conflict of interest.

References

1. Whitesides, G. The lab finally comes to the chip! *Lab Chip* **2014**, *14*, 3125–3126. [CrossRef] [PubMed]
2. Manz, A.; Graber, N.; Widmer, H.M. Miniaturized total chemical-analysis systems—A novel concept for chemical sensing. *Sens. Actuator B-Chem.* **1990**, *1*, 244–248. [CrossRef]
3. Liu, R.H.; Stremler, M.A.; Sharp, K.V.; Olsen, M.G.; Santiago, J.G.; Adrian, R.J.; Aref, H.; Beebe, D.J. Passive mixing in a three-dimensional serpentine microchannel. *J. Microelectromech. Syst.* **2000**, *9*, 190–197. [CrossRef]
4. Tekin, H.C.; Sivagnanam, V.; Ciftlik, A.T.; Sayah, A.; Vandevyver, C.; Gijs, M.A.M. Chaotic mixing using source–sink microfluidic flows in a pdms chip. *Microfluid. Nanofluid.* **2010**, *10*, 749–759. [CrossRef]
5. Wang, S.; Huang, X.; Yang, C. Mixing enhancement for high viscous fluids in a microfluidic chamber. *Lab Chip* **2011**, *11*, 2081–2087. [CrossRef] [PubMed]
6. Phan, H.V.; Coskun, M.B.; Sesen, M.; Pandraud, G.; Neild, A.; Alan, T. Vibrating membrane with discontinuities for rapid and efficient microfluidic mixing. *Lab Chip* **2015**, *15*, 4206–4216. [CrossRef] [PubMed]
7. Du, M.; Ma, Z.; Ye, X.; Zhou, Z. On-chip fast mixing by a rotary peristaltic micropump with a single structural layer. *Sci. China Technol. Sci.* **2013**, *56*, 1047–1054. [CrossRef]
8. Nguyen, N.-T.; Wu, Z. Micromixers—A review. *J. Micromech. Microeng.* **2005**, *15*, R1–R16. [CrossRef]
9. Ward, K.; Fan, Z.H. Mixing in microfluidic devices and enhancement methods. *J. Micromech. Microeng.* **2015**, *25*, 094001. [CrossRef]
10. Hessel, V.; Löwe, H.; Schönfeld, F. Micromixers—A review on passive and active mixing principles. *Chem. Eng. Sci.* **2005**, *60*, 2479–2501. [CrossRef]
11. Lee, C.Y.; Chang, C.L.; Wang, Y.N.; Fu, L.M. Microfluidic mixing: A review. *Int. J. Mol. Sci.* **2011**, *12*, 3263–3287. [CrossRef] [PubMed]
12. Lee, C.Y.; Wang, W.T.; Liu, C.C.; Fu, L.M. Passive mixers in microfluidic systems: A review. *Chem. Eng. J.* **2016**, *288*, 146–160. [CrossRef]
13. Wang, X.; Ma, X.; An, L.; Kong, X.; Xu, Z.; Wang, J. A pneumatic micromixer facilitating fluid mixing at a wide range flow rate for the preparation of quantum dots. *Sci. China Chem.* **2012**, *56*, 799–805. [CrossRef]

14. Abbas, Y.; Miwa, J.; Zengerle, R.; von Stetten, F. Active continuous-flow micromixer using an external braille pin actuator array. *Micromachines* **2013**, *4*, 80–89. [CrossRef]

15. Xia, Q.; Zhong, S. Liquid mixing enhanced by pulse width modulation in a y-shaped jet configuration. *Fluid Dyn. Res.* **2013**, *45*, 025504. [CrossRef]

16. Cortes-Quiroz, C.A.; Azarbadegan, A.; Johnston, I.D.; Tracey, M.C. Analysis and design optimization of an integrated micropump-micromixer operated for bio-mems applications. In Proceedings of the 4th Micro and Nano Flows Conference, London, UK, 7–10 September 2014; pp. 261–271.

17. Huang, J.J.; Lo, Y.J.; Hsieh, C.M.; Lei, U. An electro-thermal micro mixer. In Proceedings of the IEEE International Conference on Nano/micro Engineered and Molecular Systems, Kaohsiung, Taiwan, 20–23 February 2011; pp. 919–922.

18. Kumar, V.; Paraschivoiu, M.; Nigam, K.D.P. Single-phase fluid flow and mixing in microchannels. *Chem. Eng. Sci.* **2011**, *66*, 1329–1373. [CrossRef]

19. Posner, J.D.; Perez, C.L.; Santiago, J.G. Electric fields yield chaos in microflows. *Proc. Natl. Acad. Sci. USA* **2012**, *109*, 14353–14356. [CrossRef] [PubMed]

20. Jalaal, M.; Khorshidi, B.; Esmaeilzadeh, E. Electrohydrodynamic (EHD) mixing of two miscible dielectric liquids. *Chem. Eng. J.* **2013**, *219*, 118–123. [CrossRef]

21. Wang, G.R.; Yang, F.; Zhao, W. There can be turbulence in microfluidics at low reynolds number. *Lab Chip* **2014**, *14*, 1452–1458. [CrossRef] [PubMed]

22. Eribol, P.; Uguz, A.K. Experimental investigation of electrohydrodynamic instabilities in micro channels. *Eur. Phys. J. Spec. Top.* **2015**, *224*, 425–434. [CrossRef]

23. Thanjavur Kumar, D.; Zhou, Y.; Brown, V.; Lu, X.; Kale, A.; Yu, L.; Xuan, X. Electric field-induced instabilities in ferrofluid microflows. *Microfluid. Nanofluid.* **2015**, *19*, 43–52. [CrossRef]

24. Lang, Q.; Ren, Y.; Hobson, D.; Tao, Y.; Hou, L.; Jia, Y.; Hu, Q.; Liu, J.; Zhao, X.; Jiang, H. In-plane microvortices micromixer-based AC electrothermal for testing drug induced death of tumor cells. *Biomicrofluidics* **2016**, *10*, 064102. [CrossRef] [PubMed]

25. Wu, Y.; Ren, Y.; Tao, Y.; Hou, L.; Hu, Q.; Jiang, H. A novel micromixer based on the alternating current-flow field effect transistor. *Lab Chip* **2016**, *17*, 186–197. [CrossRef] [PubMed]

26. Zhou, T.; Wang, H.; Shi, L.; Liu, Z.; Joo, S. An enhanced electroosmotic micromixer with an efficient asymmetric lateral structure. *Micromachines* **2016**, *7*, 218. [CrossRef]

27. Baygents, J.C.; Baldessari, F. Electrohydrodynamic instability in a thin fluid layer with an electrical conductivity gradient. *Phys. Fluids* **1998**, *10*, 301–311. [CrossRef]

28. Oddy, M.H.; Santiago, J.G.; Mikkelsen, J.C. Electrokinetic instability micromixing. *Anal. Chem.* **2001**, *73*, 5822–5832. [CrossRef] [PubMed]

29. Chen, C.-H.; Lin, H.A.O.; Lele, S.K.; Santiago, J.G. Convective and absolute electrokinetic instability with conductivity gradients. *J. Fluid Mech.* **2005**, *524*, 263–303. [CrossRef]

30. Posner, J.D.; Santiago, J.G. Convective instability of electrokinetic flows in a cross-shaped microchannel. *J. Fluid Mech.* **2006**, *555*, 1. [CrossRef]

31. Boy, D.A.; Storey, B.D. Electrohydrodynamic instabilities in microchannels with time periodic forcing. *Phys. Rev. E* **2007**, *76*, 026304. [CrossRef] [PubMed]

32. Chang, C.-C.; Yang, R.-J. Electrokinetic mixing in microfluidic systems. *Microfluid. Nanofluid.* **2007**, *3*, 501–525. [CrossRef]

33. Mansur, E.A.; Mingxing, Y.E.; Yundong, W.A.N.G.; Youyuan, D.A.I. A state-of-the-art review of mixing in microfluidic mixers. *Chin. J. Chem. Eng.* **2008**, *16*, 503–516. [CrossRef]

34. Xie, Y.; Ahmed, D.; Lapsley, M.I.; Lin, S.C.; Nawaz, A.A.; Wang, L.; Huang, T.J. Single-shot characterization of enzymatic reaction constants km and kcat by an acoustic-driven, bubble-based fast micromixer. *Anal. Chem.* **2012**, *84*, 7495–7501. [CrossRef] [PubMed]

35. Xu, L.; Lee, H.; Panchapakesan, R.; Oh, K.W. Fusion and sorting of two parallel trains of droplets using a railroad-like channel network and guiding tracks. *Lab Chip* **2012**, *12*, 3936–3942. [CrossRef] [PubMed]

36. Huang, P.H.; Xie, Y.; Ahmed, D.; Rufo, J.; Nama, N.; Chen, Y.; Chan, C.Y.; Huang, T.J. An acoustofluidic micromixer based on oscillating sidewall sharp-edges. *Lab Chip* **2013**, *13*, 3847–3852. [CrossRef] [PubMed]

37. Wang, C.; Rallabandi, B.; Hilgenfeldt, S. Frequency dependence and frequency control of microbubble streaming flows. *Phys. Fluids* **2013**, *25*, 022002. [CrossRef]

38. Ozcelik, A.; Ahmed, D.; Xie, Y.; Nama, N.; Qu, Z.; Nawaz, A.A.; Huang, T.J. An acoustofluidic micromixer via bubble inception and cavitation from microchannel sidewalls. *Anal. Chem.* **2014**, *86*, 5083–5088. [CrossRef] [PubMed]

39. Brandhoff, L.; Zirath, H.; Salas, M.; Haller, A.; Peham, J.; Wiesinger-Mayr, H.; Spittler, A.; Schnetz, G.; Lang, W.; Vellekoop, M.J. A multi-purpose ultrasonic streaming mixer for integrated magnetic bead elisas. *J. Micromech. Microeng.* **2015**, *25*, 104001. [CrossRef]

40. Van 't Oever, J.; Spannenburg, N.; Offerhaus, H.; van den Ende, D.; Herek, J.; Mugele, F. In-chip direct laser writing of a centimeter-scale acoustic micromixer. *J. Micro/Nanolithogr. MEMS MOEMS* **2015**, *14*, 023503. [CrossRef]

41. Ang, K.M.; Yeo, L.Y.; Hung, Y.M.; Tan, M.K. Amplitude modulation schemes for enhancing acoustically-driven microcentrifugation and micromixing. *Biomicrofluidics* **2016**, *10*, 054106. [CrossRef] [PubMed]

42. Nama, N.; Huang, P.H.; Huang, T.J.; Costanzo, F. Investigation of micromixing by acoustically oscillated sharp-edges. *Biomicrofluidics* **2016**, *10*, 024124. [CrossRef] [PubMed]

43. Orbay, S.; Ozcelik, A.; Lata, J.; Kaynak, M.; Wu, M.; Huang, T.J. Mixing high-viscosity fluids via acoustically driven bubbles. *J. Micromech. Microeng.* **2017**, *27*, 015008. [CrossRef]

44. Eickenberg, B.; Wittbracht, F.; Stohmann, P.; Schubert, J.R.; Brill, C.; Weddemann, A.; Hütten, A. Continuous-flow particle guiding based on dipolar coupled magnetic superstructures in rotating magnetic fields. *Lab Chip* **2013**, *13*, 920. [CrossRef] [PubMed]

45. Gray, B.L.; Becker, H.; Owen, D.; Ballard, M.; Mao, W.; Alexeev, A.; Hesketh, P.J. Magnetic microbeads for sampling and mixing in a microchannel. *Int. Soc. Opt. Photonics* **2014**, *8976*, 89760C. [CrossRef]

46. Köhler, J.; Ghadiri, R.; Ksouri, S.I.; Guo, Q.; Gurevich, E.L.; Ostendorf, A. Generation of microfluidic flow using an optically assembled and magnetically driven microrotor. *J. Phys. D: Appl. Phys.* **2014**, *47*, 505501. [CrossRef]

47. La, M.; Kim, W.; Yang, W.; Kim, H.W.; Kim, D.S. Design and numerical simulation of complex flow generation in a microchannel by magnetohydrodynamic (mhd) actuation. *Int. J. Precis. Eng. Manuf.* **2014**, *15*, 463–470. [CrossRef]

48. Cao, Q.; Han, X.; Li, L. An active microfluidic mixer utilizing a hybrid gradient magnetic field. *Int. J. Appl. Electromagn. Mech.* **2015**, *47*, 583–592.

49. Kitenbergs, G.; Erglis, K.; Perzynski, R.; Cēbers, A. Magnetic particle mixing with magnetic micro-convection for microfluidics. *J. Magn. Magn. Mater.* **2015**, *380*, 227–230. [CrossRef]

50. Veldurthi, N.; Chandel, S.; Bhave, T.; Bodas, D. Computational fluid dynamic analysis of poly(dimethyl siloxane) magnetic actuator based micromixer. *Sens. Actuators B Chem.* **2015**, *212*, 419–424. [CrossRef]

51. Ballard, M.; Owen, D.; Mills, Z.G.; Hesketh, P.J.; Alexeev, A. Orbiting magnetic microbeads enable rapid microfluidic mixing. *Microfluid. Nanofluid.* **2016**, *20*, 88. [CrossRef]

52. Chang, M.; Gabayno, J.L.F.; Ye, R.; Huang, K.-W.; Chang, Y.-J. Mixing efficiency enhancing in micromixer by controlled magnetic stirring of fe3o4 nanomaterial. *Microsyst. Technol.* **2016**, *23*, 457–463. [CrossRef]

53. Hejazian, M.; Phan, D.-T.; Nguyen, N.-T. Mass transport improvement in microscale using diluted ferrofluid and a non-uniform magnetic field. *RSC Adv.* **2016**, *6*, 62439–62444. [CrossRef]

54. Owen, D.; Ballard, M.; Alexeev, A.; Hesketh, P.J. Rapid microfluidic mixing via rotating magnetic microbeads. *Sens. Actuators A: Phys.* **2016**, *251*, 84–91. [CrossRef]

55. Veldurthi, N.; Ghoderao, P.; Sahare, S.; Kumar, V.; Bodas, D.; Kulkarni, A.; Bhave, T. Magnetically active micromixer assisted synthesis of drug nanocomplexes exhibiting strong bactericidal potential. *Mater. Sci. Eng. C* **2016**, *68*, 455–464. [CrossRef] [PubMed]

56. Hejazian, M.; Nguyen, N.-T. A rapid magnetofluidic micromixer using diluted ferrofluid. *Micromachines* **2017**, *8*, 37. [CrossRef]

57. Nouri, D.; Zabihi-Hesari, A.; Passandideh-Fard, M. Rapid mixing in micromixers using magnetic field. *Sens. Actuators A: Phys.* **2017**, *255*, 79–86. [CrossRef]

58. Dong, X.X.; Zhang, L.; Fu, J. Laser-induced thermal bubble-mixing on a microfluidic platform for lab-on-a-chip applications. *Adv. Mater. Res.* **2012**, *557–559*, 2197–2201. [CrossRef]

59. Huang, K.-R.; Chang, J.-S.; Chao, S.D.; Wung, T.-S.; Wu, K.-C. Study of active micromixer driven by electrothermal force. *Jpn. J. Appl. Phys.* **2012**, *51*, 047002. [CrossRef]

60. Sasaki, N.; Kitamori, T.; Kim, H.B. Fluid mixing using ac electrothermal flow on meandering electrodes in a microchannel. *Electrophoresis* **2012**, *33*, 2668–2673. [CrossRef] [PubMed]

61. Huang, C.; Tsou, C. The implementation of a thermal bubble actuated microfluidic chip with microvalve, micropump and micromixer. *Sens. Actuators A: Phys.* **2014**, *210*, 147–156. [CrossRef]

62. Zhang, F.; Chen, H.; Chen, B.; Wu, J. Alternating current electrothermal micromixer with thin film resistive heaters. *Adv. Mech. Eng.* **2016**, *8*, 168781401664626. [CrossRef]

63. Kunti, G.; Bhattacharya, A.; Chakraborty, S. Rapid mixing with high-throughput in a semi-active semi-passive micromixer. *Electrophoresis* **2017**, *38*, 1310–1317. [CrossRef] [PubMed]

64. Glasgow, I.; Aubry, N. Enhancement of microfluidic mixing using time pulsing. *Lab Chip* **2003**, *3*, 114–120. [CrossRef] [PubMed]

65. Niu, X.Z.; Lee, Y.K. Efficient spatial-temporal chaotic mixing in microchannels. *J. Micromech. Microeng.* **2003**, *13*, 454–462. [CrossRef]

66. Tabeling, P.; Chabert, M.; Dodge, A.; Jullien, C.; Okkels, F. Chaotic mixing in cross-channel micromixers. *Philos. T. R. Soc. A* **2004**, *362*, 987–1000. [CrossRef] [PubMed]

67. Deshmukh, A.A.; Liepmann, D.; Pisano, A.P. Continuous micromixer with pulsatile micropumps. In Proceedings of the Technical Digest of the IEEE Solid State Sensor and Actuator Workshop, Hilton Head Island, SC, USA, January 2000; pp. 73–76.

68. Xia, Q.; Zhong, S. Quantification of liquid mixing enhanced by alternatively pulsed injection in a confined jet configuration. *J. Vis.* **2011**, *15*, 57–66. [CrossRef]

69. Khoshmanesh, K.; Almansouri, A.; Albloushi, H.; Yi, P.; Soffe, R.; Kalantar-zadeh, K. A multi-functional bubble-based microfluidic system. *Sci. Rep.* **2015**, *5*, 9942. [CrossRef] [PubMed]

70. Sun, C.-L.; Sie, J.-Y. Active mixing in diverging microchannels. *Microfluid. Nanofluid.* **2009**, *8*, 485–495. [CrossRef]

71. Yi-Kuen, L.; Deval, J.; Tabeling, P.; Chih-Ming, H. Chaotic mixing in electrokinetically and pressure driven micro flows. In Proceedings of the 14th IEEE International Conference on Micro Electro Mechanical Systems, Interlaken, Switzerland, 25 January 2001; pp. 483–486. [CrossRef]

72. Lin, H. Electrokinetic instability in microchannel flows: A review. *Mech. Res. Commun.* **2009**, *36*, 33–38. [CrossRef]

73. Tang, S.-Y.; Sivan, V.; Petersen, P.; Zhang, W.; Morrison, P.D.; Kalantar-zadeh, K.; Mitchell, A.; Khoshmanesh, K. Liquid metal actuator for inducing chaotic advection. *Adv. Funct. Mater.* **2014**, *24*, 5851–5858. [CrossRef]

74. Tang, S.Y.; Khoshmanesh, K.; Sivan, V.; Petersen, P.; O'Mullane, A.P.; Abbott, D.; Mitchell, A.; Kalantar-zadeh, K. Liquid metal enabled pump. *Proc. Natl. Acad. Sci. USA* **2014**, *111*, 3304–3309. [CrossRef] [PubMed]

75. Moroney, R.M.; White, R.M.; Howe, R.T. Ultrasonically induced microtransport. In Proceedings of the IEEE Micro Electro Mechanical Systems, Nara, Japan, 1–2 January 1991; pp. 277–282.

76. Hashmi, A.; Yu, G.; Reilly-Collette, M.; Heiman, G.; Xu, J. Oscillating bubbles: A versatile tool for lab on a chip applications. *Lab Chip* **2012**, *12*, 4216–4227. [CrossRef] [PubMed]

77. Liu, R.H.; Yang, J.; Pindera, M.Z.; Athavale, M.; Grodzinski, P. Bubble-induced acoustic micromixing. *Lab Chip* **2002**, *2*, 151–157. [CrossRef] [PubMed]

78. Liu, R.H.; Lenigk, R.; Druyor-Sanchez, R.L.; Yang, J.N.; Grodzinski, P. Hybridization enhancement using cavitation microstreaming. *Anal. Chem.* **2003**, *75*, 1911–1917. [CrossRef] [PubMed]

79. Ahmed, D.; Mao, X.; Shi, J.; Juluri, B.K.; Huang, T.J. A millisecond micromixer via single-bubble-based acoustic streaming. *Lab Chip* **2009**, *9*, 2738–2741. [CrossRef] [PubMed]

80. Wang, S.; Huang, X.; Yang, C. Microfluidic bubble generation by acoustic field for mixing enhancement. *J. Heat Transf.* **2012**, *134*, 051014. [CrossRef]

81. Meng, L.; Cai, F.; Jin, Q.; Niu, L.; Jiang, C.; Wang, Z.; Wu, J.; Zheng, H. Acoustic aligning and trapping of microbubbles in an enclosed pdms microfluidic device. *Sens. Actuators B Chem.* **2011**, *160*, 1599–1605. [CrossRef]

82. Luong, T.D.; Phan, V.N.; Nguyen, N.T. High-throughput micromixers based on acoustic streaming induced by surface acoustic wave. *Microfluid. Nanofluid.* **2011**, *10*, 619–625. [CrossRef]

83. Qian, S.; Bau, H.H. Magneto-hydrodynamic stirrer for stationary and moving fluids. *Sens. Actuat B-Chem.* **2005**, *106*, 859–870. [CrossRef]

84. Ryu, K.S.; Shaikh, K.; Goluch, E.; Fan, Z.; Liu, C. Micro magnetic stir-bar mixer integrated with parylene microfluidic channels. *Lab Chip* **2004**, *4*, 608–613. [CrossRef] [PubMed]

85. Chen, C.Y.; Chen, C.Y.; Lin, C.Y.; Hu, Y.T. Magnetically actuated artificial cilia for optimum mixing performance in microfluidics. *Lab Chip* **2013**, *13*, 2834–2839. [CrossRef] [PubMed]

86. Tsou, C.; Huang, C. Thermal bubble microfluidic gate based on SOI wafer. *J. Microelectromech. Syst.* **2009**, *18*, 852–859. [CrossRef]

87. La, M.; Park, S.J.; Kim, H.W.; Park, J.J.; Ahn, K.T.; Ryew, S.M.; Kim, D.S. A centrifugal force-based serpentine micromixer (CSM) on a plastic lab-on-a-disk for biochemical assays. *Microfluid. Nanofluid.* **2012**, *15*, 87–98. [CrossRef]

88. Aguirre, G.R.; Efremov, V.; Kitsara, M.; Ducrée, J. Integrated micromixer for incubation and separation of cancer cells on a centrifugal platform using inertial and dean forces. *Microfluid. Nanofluid.* **2014**, *18*, 513–526. [CrossRef]

89. Shamloo, A.; Madadelahi, M.; Akbari, A. Numerical simulation of centrifugal serpentine micromixers and analyzing mixing quality parameters. *Chem. Eng. Process.: Process Intensif.* **2016**, *104*, 243–252. [CrossRef]

90. Haeberle, S.; Brenner, T.; Schlosser, H.P.; Zengerle, R.; Ducrée, J. Centrifugal micromixer. *Chem. Eng. Technol.* **2005**, *28*, 613–616. [CrossRef]

91. Leung, W.W.-F.; Ren, Y. Crossflow and mixing in obstructed and width-constricted rotating radial microchannel. *Int. J. Heat Mass Transf.* **2013**, *64*, 457–467. [CrossRef]

92. Ansari, M.A.; Kim, K.-Y. Mixing performance of unbalanced split and recombine micomixers with circular and rhombic sub-channels. *Chem. Eng. J.* **2010**, *162*, 760–767. [CrossRef]

93. Kamholz, A.E.; Weigl, B.H.; Finlayson, B.A.; Yager, P. Quantitative analysis of molecular interaction in a microfluidic channel: The t-sensor. *Anal. Chem.* **1999**, *71*, 5340. [CrossRef] [PubMed]

94. Mengeaud, V.; Josserand, J.; Girault, H.H. Mixing processes in a zigzag microchannel: Finite element simulations and optical study. *Anal. Chem.* **2002**, *74*, 4279–4286. [CrossRef] [PubMed]

95. Hossain, S.; Kim, K.-Y. Mixing analysis of passive micromixer with unbalanced three-split rhombic sub-channels. *Micromachines* **2014**, *5*, 913–928. [CrossRef]

96. Li, J.; Xia, G.; Li, Y. Numerical and experimental analyses of planar asymmetric split-and-recombine micromixer with dislocation sub-channels. *J. Chem. Technol. Biotechnol.* **2013**, *88*, 1757–1765. [CrossRef]

97. Wang, L.; Ma, S.; Wang, X.; Bi, H.; Han, X. Mixing enhancement of a passive microfluidic mixer containing triangle baffles. *Asia-Pac. J. Chem. Eng.* **2014**, *9*, 877–885. [CrossRef]

98. Alam, A.; Afzal, A.; Kim, K.-Y. Mixing performance of a planar micromixer with circular obstructions in a curved microchannel. *Chem. Eng. Res. Des.* **2014**, *92*, 423–434. [CrossRef]

99. Scherr, T.; Quitadamo, C.; Tesvich, P.; Park, D.S.; Tiersch, T.; Hayes, D.; Choi, J.W.; Nandakumar, K.; Monroe, W.T. A planar microfluidic mixer based on logarithmic spirals. *J. Micromech. Microeng.* **2012**, *22*, 55019. [CrossRef] [PubMed]

100. He, X.; Wei, D.; Deng, Z.; Yang, S.; Cai, S. Mixing performance of a novel passive micromixer with logarithmic spiral channel. *Paiguan Jixie Gongcheng Xuebao/J. Drain. Irrig. Mach. Eng.* **2014**, *32*, 968–972.

101. Afzal, A.; Kim, K.-Y. Multi-objective optimization of a passive micromixer based on periodic variation of velocity profile. *Chem. Eng. Commun.* **2014**, *202*, 322–331. [CrossRef]

102. Wu, C.-Y.; Tsai, R.-T. Fluid mixing via multidirectional vortices in converging–diverging meandering microchannels with semi-elliptical side walls. *Chem. Eng. J.* **2013**, *217*, 320–328. [CrossRef]

103. Afzal, A.; Kim, K.-Y. Passive split and recombination micromixer with convergent–divergent walls. *Chem. Eng. J.* **2012**, *203*, 182–192. [CrossRef]

104. Tran-Minh, N.; Dong, T.; Karlsen, F. An efficient passive planar micromixer with ellipse-like micropillars for continuous mixing of human blood. *Comput. Methods Programs Biomed.* **2014**, *117*, 20–29. [CrossRef] [PubMed]

105. Chen, X.; Li, T. A novel design for passive misscromixers based on topology optimization method. *Biomed. Microdevices* **2016**, *18*, 57. [CrossRef] [PubMed]

106. Chen, X.; Li, T. A novel passive micromixer designed by applying an optimization algorithm to the zigzag microchannel. *Chem. Eng. J.* **2017**, *313*, 1406–1414. [CrossRef]

107. The, H.L.; Le-Thanh, H.; Tran-Minh, N.; Karlsen, F. A novel passive micromixer with trapezoidal blades for high mixing efficiency at low reynolds number flow. In Proceedings of the 2014 Middle East Conference on Biomedical Engineering (MECBME), Doha, Qatar, 17–20 February 2014; pp. 25–28.

108. The, H.L.; Ta, B.Q.; Thanh, H.L.; Dong, T.; Thoi, T.N.; Karlsen, F. Geometric effects on mixing performance in a novel passive micromixer with trapezoidal-zigzag channels. *J. Micromech. Microeng.* **2015**, *25*, 094004. [CrossRef]

109. Le The, H.; Tran-Minh, N.; Le-Thanh, H.; Karlsen, F. A novel micromixer with multimixing mechanisms for high mixing efficiency at low reynolds number. In Proceedings of the 2014 9th Ieee International Conference on Nano/Micro Engineered and Molecular Systems (NEMS), Waikiki Beach, HI, USA, 13–16 April 2014; pp. 653–656.

110. Alam, A.; Kim, K.-Y. Mixing performance of a planar micromixer with circular chambers and crossing constriction channels. *Sens. Actuators B Chem.* **2013**, *176*, 639–652. [CrossRef]

111. Viktorov, V.; Mahmud, M.R.; Visconte, C. Numerical study of fluid mixing at different inlet flow-rate ratios in tear-drop and chain micromixers compared to a new h-c passive micromixer. *Eng. Appl. Comput. Fluid Mech.* **2016**, *10*, 182–192. [CrossRef]

112. Yang, J.; Qi, L.; Chen, Y.; Ma, H. Design and fabrication of a three dimensional spiral micromixer. *Chin. J. Chem.* **2013**, *31*, 209–214. [CrossRef]

113. Liu, K.; Yang, Q.; Chen, F.; Zhao, Y.; Meng, X.; Shan, C.; Li, Y. Design and analysis of the cross-linked dual helical micromixer for rapid mixing at low reynolds numbers. *Microfluid. Nanofluid.* **2015**, *19*, 169–180. [CrossRef]

114. Sheu, T.S.; Chen, S.J.; Chen, J.J. Mixing of a split and recombine micromixer with tapered curved microchannels. *Chem. Eng. Sci.* **2012**, *71*, 321–332. [CrossRef]

115. Li, X.; Chang, H.; Liu, X.; Ye, F.; Yuan, W. A 3-d overbridge-shaped micromixer for fast mixing over a wide range of reynolds numbers. *J. Microelectromech. Syst.* **2015**, *24*, 1391–1399. [CrossRef]

116. Yang, A.-S.; Chuang, F.-C.; Chen, C.-K.; Lee, M.-H.; Chen, S.-W.; Su, T.-L.; Yang, Y.-C. A high-performance micromixer using three-dimensional tesla structures for bio-applications. *Chem. Eng. J.* **2015**, *263*, 444–451. [CrossRef]

117. Feng, X.; Ren, Y.; Jiang, H. An effective splitting-and-recombination micromixer with self-rotated contact surface for wide reynolds number range applications. *Biomicrofluidics* **2013**, *7*, 54121. [CrossRef] [PubMed]

118. Hossain, S.; Lee, I.; Kim, S.M.; Kim, K.-Y. A micromixer with two-layer serpentine crossing channels having excellent mixing performance at low reynolds numbers. *Chem. Eng. J.* **2017**, *327*, 268–277. [CrossRef]

119. Stroock, A.D.; Dertinger, S.K.; Ajdari, A.; Mezic, I.; Stone, H.A.; Whitesides, G.M. Chaotic mixer for microchannels. *Science* **2002**, *295*, 647–651. [CrossRef] [PubMed]

120. Howell, P.B., Jr.; Mott, D.R.; Fertig, S.; Kaplan, C.R.; Golden, J.P.; Oran, E.S.; Ligler, F.S. A microfluidic mixer with grooves placed on the top and bottom of the channel. *Lab Chip* **2005**, *5*, 524–530. [CrossRef] [PubMed]

121. Hossain, S.; Husain, A.; Kim, K.-Y. Optimization of micromixer with staggered herringbone grooves on top and bottom walls. *Eng. Appl. Comput. Fluid Mech.* **2014**, *5*, 506–516. [CrossRef]

122. Bhagat, A.A.S.; Peterson, E.T.K.; Papautsky, I. A passive planar micromixer with obstructions for mixing at low reynolds numbers. *J. Micromech. Microeng.* **2007**, *17*, 1017–1024. [CrossRef]

123. Sadegh Cheri, M.; Latifi, H.; Salehi Moghaddam, M.; Shahraki, H. Simulation and experimental investigation of planar micromixers with short-mixing-length. *Chem. Eng. J.* **2013**, *234*, 247–255. [CrossRef]

124. Lee, C.Y.; Lin, C.F.; Hung, M.F.; Ma, R.H.; Tsai, C.H.; Lin, C.H.; Fu, L.M. Experimental and numerical investigation into mixing efficiency of micromixers with different geometric barriers. *Mater. Sci. Forum* **2006**, *505–507*, 391–396. [CrossRef]

125. Hossain, S.; Ansari, M.A.; Kim, K.-Y. Evaluation of the mixing performance of three passive micromixers. *Chem. Eng. J.* **2009**, *150*, 492–501. [CrossRef]

126. Jiang, F.; Drese, K.S.; Hardt, S.; Küpper, M.; Schönfeld, F. Helical flows and chaotic mixing in curved micro channels. *AIChE J.* **2004**, *50*, 2297–2305. [CrossRef]

127. Sudarsan, A.P.; Ugaz, V.M. Multivortex micromixing. *Proc. Natl. Acad. Sci. USA* **2006**, *103*, 7228–7233. [CrossRef] [PubMed]

128. Tsai, R.T.; Wu, C.Y. An efficient micromixer based on multidirectional vortices due to baffles and channel curvature. *Biomicrofluidics* **2011**, *5*, 14103. [CrossRef] [PubMed]

129. Dean, W.R. Fluid motion in a curved channel. *Proc. R. Soc. A: Math. Phys. Eng. Sci.* **1928**, *121*, 402–420. [CrossRef]

130. Ansari, M.A.; Kim, K.-Y.; Anwar, K.; Kim, S.M. A novel passive micromixer based on unbalanced splits and collisions of fluid streams. *J. Micromech. Microeng.* **2010**, *20*, 055007. [CrossRef]

131. Xia, G.; Li, J.; Wu, H.; Zhou, M. Mixing performance of asymmetric split and recombine micromixer with fan-shaped cavity. *CIESC J.* **2011**, *62*, 1219–1225.
132. Schönfeld, F.; Hardt, S. Simulation of helical flows in microchannels. *AIChE J.* **2004**, *50*, 771–778. [CrossRef]
133. Sudarsan, A.P.; Ugaz, V.M. Fluid mixing in planar spiral microchannels. *Lab Chip* **2006**, *6*, 74–82. [CrossRef] [PubMed]
134. Li, P.; Cogswell, J.; Faghri, M. Design and test of a passive planar labyrinth micromixer for rapid fluid mixing. *Sens. Actuators B Chem.* **2012**, *174*, 126–132. [CrossRef]
135. Al-Halhouli, A.A.; Alshare, A.; Mohsen, M.; Matar, M.; Dietzel, A.; Büttgenbach, S. Passive micromixers with interlocking semi-circle and omega-shaped modules: Experiments and simulations. *Micromachines* **2015**, *6*, 953–968. [CrossRef]
136. Afzal, A.; Kim, K.-Y. Convergent–divergent micromixer coupled with pulsatile flow. *Sens. Actuators B Chem.* **2015**, *211*, 198–205. [CrossRef]
137. Yakhshi Tafti, E.; Kumar, R.; Cho, H.J. Effect of laminar velocity profile variation on mixing in microfluidic devices: The sigma micromixer. *Appl. Phys. Lett.* **2008**, *93*, 143504. [CrossRef]
138. Branebjerg, J.; Gravesen, P.; Krog, J.P.; Nielsen, C.R. Fast mixing by lamination. In Proceedings of the Micro Electro Mechanical Systems, San Diego, CA, USA, 11–15 February 1996; pp. 441–446. [CrossRef]
139. Buchegger, W.; Wagner, C.; Lendl, B.; Kraft, M.; Vellekoop, M.J. A highly uniform lamination micromixer with wedge shaped inlet channels for time resolved infrared spectroscopy. *Microfluid. Nanofluid.* **2010**, *10*, 889–897. [CrossRef]
140. SadAbadi, H.; Packirisamy, M.; Wüthrich, R. High performance cascaded pdms micromixer based on split-and-recombination flows for lab-on-a-chip applications. *RSC Adv.* **2013**, *3*, 7296. [CrossRef]
141. Lim, T.W.; Son, Y.; Jeong, Y.J.; Yang, D.Y.; Kong, H.J.; Lee, K.S.; Kim, D.P. Three-dimensionally crossing manifold micro-mixer for fast mixing in a short channel length. *Lab Chip* **2011**, *11*, 100–103. [CrossRef] [PubMed]
142. Nimafar, M.; Viktorov, V.; Martinelli, M. Experimental comparative mixing performance of passive micromixers with h-shaped sub-channels. *Chem. Eng. Sci.* **2012**, *76*, 37–44. [CrossRef]
143. Viktorov, V.; Nimafar, M. A novel generation of 3d sar-based passive micromixer: Efficient mixing and low pressure drop at a low reynolds number. *J. Micromech. Microeng.* **2013**, *23*, 055023. [CrossRef]
144. Le The, H.; Le Thanh, H.; Dong, T.; Ta, B.Q.; Tran-Minh, N.; Karlsen, F. An effective passive micromixer with shifted trapezoidal blades using wide reynolds number range. *Chem. Eng. Res. Des.* **2015**, *93*, 1–11. [CrossRef]
145. And, N.S.; Tafti, D.K. Evaluation of microchamber geometries and surface conditions for electrokinetic driven mixing. *Anal. Chem.* **2004**, *76*, 3785.
146. Böhm, S.; Greiner, K.; Schlautmann, S.; Vries, S.D.; Berg, A.V.D. A rapid vortex micromixer for studying high-speed chemical reactions. In *Micro Total Analysis Systems 2001*; Springer: Monterey, CA, USA, 21–25 October 2001.
147. Chung, Y.C.; Hsu, Y.L.; Jen, C.P.; Lu, M.C.; Lin, Y.C. Design of passive mixers utilizing microfluidic self-circulation in the mixing chamber. *Lab Chip* **2003**, *4*, 70–77. [CrossRef] [PubMed]
148. Rafeie, M.; Welleweerd, M.; Hassanzadeh-Barforoushi, A.; Asadnia, M.; Olthuis, W.; Ebrahimi Warkiani, M. An easily fabricated three-dimensional threaded lemniscate-shaped micromixer for a wide range of flow rates. *Biomicrofluidics* **2017**, *11*, 014108. [CrossRef] [PubMed]
149. Gong, H.; Woolley, A.T.; Nordin, G.P. High density 3d printed microfluidic valves, pumps, and multiplexers. *Lab Chip* **2016**, *16*, 2450–2458. [CrossRef] [PubMed]

MDPI

St. Alban-Anlage 66

4052 Basel

Switzerland

Tel. +41 61 683 77 34

Fax +41 61 302 89 18

www.mdpi.com

Actuators Editorial Office

E-mail: actuators@mdpi.com

www.mdpi.com/journal/actuators

www.ingramcontent.com/pod-product-compliance
Lightning Source LLC
Chambersburg PA
CBHW051900210326
41597CB00033B/5970